FLORA of the MOUNT HAMILTON RANGE

California Native Plant Society

FLORA

of the

Mount Hamilton Range of California

Helen K. Sharsmith

with index compiled by
Carl W. Sharsmith
and
Nobi Kurotori

California Native Plant Society

Special Publication Number 6

The original article *"Flora of the Mount Hamilton Range
of California"* appeared in *The American Midland
Naturalist* Vol. 34, No. 2 (Sept. 1945), copyright held by
University of Notre Dame, reproduced here by permission.
Artwork from *Spring Wildflowers of the San Francisco Bay
Region* by Helen K. Sharsmith is used here by permission
of the publisher, University of California Press.

Publication coordinated by June Bilisoly.
Designed and produced by Dave Comstock.

Library of Congress Cataloging in Publication Data

Sharsmith, Helen K.
  Flora of the Mount Hamilton Range of California.

  (Special publication / California Native Plant
Society; no. 6)
  Originally presented as the author's thesis
(Ph.D.—University of California, Berkeley, 1940)
  Reprint from the American midland naturalist, v. 34,
no. 2, Sept. 1945.
  Includes index.
  1. Botany — California — Hamilton Range, Mount.
2. Hamilton, Range, Mount (Calif.) I. American midland
naturalist. II. Title. III. Series: Special publication
(California Native Plant Society); no. 6.
QK149.S47   1982   582.09794'73   82-9600
ISBN 0-943460-08-5

Foreword

It is a privilege to be invited by the Santa Clara Chapter of the California Native Plant Society to write the Foreword for their republication of Helen Sharsmith's *Flora of the Mount Hamilton Range of California.* The manuscript of this Flora served as the thesis for her Ph.D. degree, awarded by the University of California, Berkeley, in 1940. Fortunately, civilization has not yet encroached on the Mount Hamilton Range to such an extent that the flora has been much altered. Most of the species, some of which are now considered endangered, may still be found there.

During the period that she was carrying on field studies in the Mount Hamilton Range, it was my good fortune to be Helen's companion on many weekend field trips. Together, we happily followed the tiny unimproved road along Arroyo Bayo with its innumerable wet crossings, enjoyed the delicate beauty of *Streptanthus callistus* and the exotic fragrance of *Clarkia breweri*, and we shared the excitement of finding a *Lotus* unknown to us—later published by Helen as *Lotus rubriflorus*. Sometimes, on early spring trips, we would emerge from our sleeping bags to fields white with frost; once we were caught by an unanticipated storm with resulting skiddy roads; and once, on the east side of the Range, we had the misfortune of having large, soft-backed cattle ticks as sleeping companions! Helen's intimate knowledge of and concern for plants was brought out by an incident indelibly engraved on my memory. On one of our early spring trips, with long, sliding steps I had exuberantly descended one of the steep, fine talus slopes above Colorado Creek in the Red Mountain area. Later, Helen gently, but reprovingly remarked that I had undoubtedly disturbed many of the serpentine endemics that would bloom on these slopes later in the spring; such plants as certain species of *Fritillaria, Streptanthus, and Campanula.* Ever since, I have tried to tread lightly and carefully upon Mother Earth. The *Campanula* that grows on those slopes was described in 1980 as *Campanula sharsmithiae* by Nancy Morin in her treatment of the annual California species of the genus—a fitting tribute to Helen's botanical work in the area.

Helen Sharsmith's appreciation of plants and the natural scene is evidenced in her early contributions to *Yosemite Nature Notes.* Later, in the *Mount Hamilton Flora*, with characteristic thoroughness and attention to detail, she ably discussed the geological, edaphic, and climatic factors that undoubtedly influenced the distribution of the plants in the area, as well as the floristic relationships of Mount Hamilton plants in relation to those of other areas. Taxonomy and nomenclature are not static phases of botany. It is inevitable, therefore, that some name changes have been proposed for the 761 species and varieties listed in the *Flora of the Mount Hamilton Range* since it was published 36 years ago. Such name changes have been indicated in the Index appended to the Flora.

It seems appropriate to take this opportunity to mention major botanical contributions made by Helen Sharsmith during the many years that we served together on the staff of the Herbarium of the University of California, Berkeley. Her monograph, "The genus *Hesperolinon* (Linaceae)", published in 1961 (Univ. Calif. Publ. Bot. 32) is cited by the English botanists Davies and Heywood in their important volume, *Principles of Angiosperm Taxonomy,* as "...an outstanding example of floral morphological studies carried on in connection with a taxonomic revision." Several species of *Hesperolinon* are found in the Mount Hamilton Range. Her *Spring Wild Flowers of the Bay Area,* published in 1965, continues to be a best seller in the University of California's Natural History Guide series. In leading local walks, it has always been her aim to point out the characteristic structure and relationship of flowers instead of just "spouting" scientific names. Plants should be friends, each with characters that can be recognized. She regretted that the exigencies of a thesis excluded much of such information from the *Flora of Mount Hamilton.*

Annetta Carter

March 1982

The American Midland Naturalist

Published Bi-Monthly by The University of Notre Dame, Notre Dame, Indiana

| Vol. 34 | SEPTEMBER, 1945 | No. 2 |

Flora of the Mount Hamilton Range of California

(A taxonomic study and floristic analysis of the vascular plants)

Helen K. Sharsmith

CONTENTS

Introduction 289
Analysis of the factors relating to the
    distribution and origin of the flora....293
    Physiography 293
    Geologic history 297
    Climate 299
    Soils 302
    Influence of man 306
Analysis of the vegetation 307
    Plant communities 307
    Seasonal periodicity 310
Geographical analysis of the species 311

Cosmopolitan species 311
Introduced species 312
Northern floristic element 313
Sonoran floristic element 315
    Great Basin and desert
      derivatives 316
    Cismontane southern Cali-
      fornia derivatives 319
Central Coast Range endemic
    element 321
Acknowledgments 325
References 325
Annotated list of vascular species 327

Introduction

A thorough understanding of the complex nature of the California flora awaits detailed investigations of the state's many topographic units which botanically are still largely unexplored. The Mount Hamilton Range was one of these topographic units, and it was chosen as an area for botanical research with the expectation that it would contribute materially to our knowledge of the California flora.

The geographical analysis of any flora is of value only when based upon a reasonably complete knowledge of its component plant species. In the Mount Hamilton Range, botanical exploration began with W. H. Brewer, field botanist of the California Geological Survey from 1860 to 1864, and his journal (Brewer, 1930) gives an account of the areas he penetrated. He and his party climbed Mount Hamilton on September 1, 1861, the earliest recorded ascent. In 1862 he worked along the eastern base of the range from Corral Hollow to Pacheco Pass, penetrating canyons which open into the San Joaquin Valley (Fig. 1). Some of these localities have not been visited since by botanists. The types of *Streptanthus Breweri*, *Monardella Breweri*, *Clarkia Breweri*, and *Oenothera deltoides* var. *cognata* were collected on this trip.

289

Brewer's collections are at the Gray Herbarium, but many duplicates are in California herbaria. E. L. Greene (1893a) spent a week on the summit of Mount Hamilton in 1893. His collections are in the herbarium of the University of Notre Dame. In 1903 A. D. E. Elmer collected rather extensively on the western side of the range, and also visited Arroyo Mocho, San Antonio

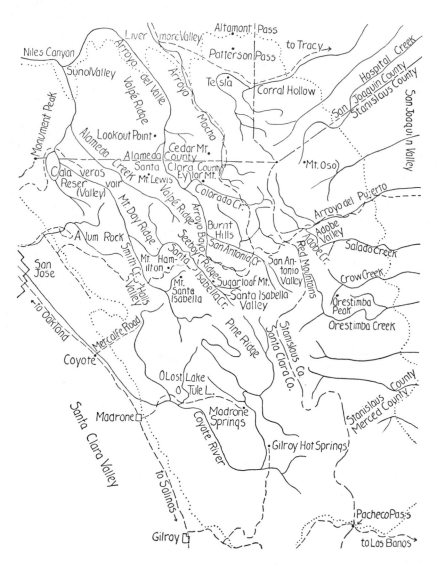

Fig. 1. Mount Hamilton Range. All collecting localities are included in this map. .... = boundaries of the range. - - - - - - = roads. ⸺ - - ⸺ = county boundaries.

Valley, Cedar Mountain, Red Mountain, Arroyo del Puerto, and Adobe Valley; his collection is at Stanford University, and duplicates are in various western herbaria. In 1907 A. A. Heller (1907) visited the Alum Rock area and Mount Hamilton. Many of his specimens are at Stanford University, the University of California, and other western herbaria. Several other early California botanists made short collecting trips to the western side of the range, among them: T. S. Brandegee (Mount Hamilton in 1890), W. R. Dudley (Pine Ridge and Gilroy Hot Springs in 1895), J. Burtt Davy (Mount Hamilton in 1897), R. J. Smith (Mount Day, Oak Ridge, Los Buellis Hills, and Mount Hamilton in 1904, 1906, and 1908), H. P. Chandler (Mount Hamilton in 1906), and R. L. Pendleton (Alum Rock and Mount Hamilton in 1907). Their collections are variously represented at the herbaria of the University of California, Stanford University, and California Academy of Sciences.

Among recent collectors, R. F. Hoover made several explorations along the eastern margin of the range from 1935 to 1938; his collections are in the University of California herbarium or his personal herbarium. The field staff of the Vegetative Type Map Herbarium of the United States Forest Service penetrated several unexplored areas between 1935 and 1938, particularly on the western side, and their collections and Hoover's gave valuable assistance in the present study. Field work was carried on by the writer from 1934 to 1937. Explorations were limited mainly to an east-west transect which included Mount Hamilton, the highest mountain in the range (Figs. 1 and 2), although many areas north and south of the transect were penetrated. This plan was chosen because 1) it embodied the major vegetational and floristic features, and 2) better knowledge of the flora was obtained by thoroughly investigating the transect area than would have been possible by a more superficial exploration of the entire range. A complete set of collected specimens was deposited at the University of California herbarium, and the duplicates were distributed from there.

There are 761 species and varieties of Mount Hamilton Range vascular plants listed in this paper. The following tabular summary indicates their taxonomic position:

	Families	Genera	Species
Pteridophyta	5	13	16
Gymnospermae	3	4	6
Monocotyledonae	10	44	110
Dicotyledonae	68	270	629
Totals	86	331	761

The three largest families are the Compositae (61 genera, 126 species), Gramineae (22 genera, 53 species), and Leguminosae (12 genera, 53 species). Other well represented families are: Scrophulariaceae, Liliaceae, Cruciferae, Hydrophyllaceae, Ranunculaceae, Umbelliferae, Polygonaceae, and Polemoniaceae.

These 761 species of vascular plants represent 19 per cent of the 4019 species listed by Jepson (1925) for the entire state of California. There are

86 families in the Mount Hamilton Range vascular flora, or 57 per cent of the 152 families which occur in the California flora. Considering the small area of the Mount Hamilton Range (approximately 1500 square miles) in relation to the area of California (158,297 square miles), the Mount Hamilton Range appears to have a well developed and diverse flora. About 47 per cent of the Mount Hamilton Range plants are entirely or almost entirely restricted to the California botanical province of Jepson (1925). This is a larger percentage of Californian species than is found within the state as a

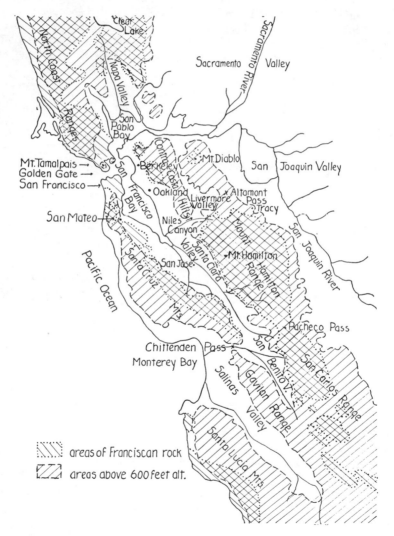

Fig. 2. Central Coast Ranges of California; showing relationship of Mount Hamilton Range to other portions of the Coast Ranges.

whole (about 40 per cent according to Jepson). The distinction and diversity of the Mount Hamilton Range flora seem to be due to the same factors which have resulted in the distinction and diversity of the California flora as a whole, namely the development or migration and establishment of diverse floristic elements in response to the combined influences of a complex geologic history and a resultant complexity of topography, climate, and soils.

Analysis of the Factors Relating to the Distribution and Origin of the Flora

PHYSIOGRAPHY

The Mount Hamilton Range is one of the several subdivisions of the Diablo Range of the inner South Coast Ranges of California. The Diablo Range (Figs. 2 and 3) extends in a northwest-southeast direction as a more or less continuous mountain chain some twenty to thirty miles wide, from San Pablo Bay in central California to Polonio Pass in northeastern San Luis Obispo County. South of Polonio Pass in the inner South Coast Ranges, the Temblor Range connects the Diablo Range with the Tehachapi Mountains of southern California, and the Tehachapi Mountains in turn link the inner South Coast Ranges with the Sierra Nevada (Anderson and Pack, 1915; Anderson, 1905; Clark, 1929). On the west the Diablo Range is bordered, from north to south, by San Francisco Bay, Santa Clara Valley, San Benito Valley, the south end of the Gavilan Range where it merges with the Diablo Range, the Salinas Valley, and Cholame Valley. On the east it is bordered for its entire length by the San Joaquin Valley. The interior of the Diablo Range has a minimum altitude of 2000 to 3000 feet throughout its length, and is broken by only four or five east-west passes. These represent major topographic breaks, dividing the Diablo Range into several more or less distinct topographic subdivisions which roughly correspond to the structural subdivisions of the range. The topographic subdivisions overlap somewhat, but in general they form northwest-southeast trending units known as Contra Costa Hills (the most northerly), Mount Diablo, Mount Hamilton Range, Panoche Hills, San Carlos Range (which is here regarded as including the Panoche Hills), and the Estrella Hills (Whitney, 1865).

The Mount Hamilton Range (Fig. 1) subdivision of the Diablo Range forms an unbroken, well defined, and relatively isolated mountain block approximately fifty miles long and thirty miles wide. It is delimited on the north by Niles (Alameda Creek) Canyon, Sunol Valley, Livermore Valley, and Altamont (Livermore) Pass (740 feet), on the south by Pacheco Pass (1470 feet), on the west by the Santa Clara Valley, and on the east by the San Joaquin Valley. The Altamont and Pacheco passes represent major topographic breaks in the Diablo Range as a whole. Lawson and Palach (1902) consider the Contra Costa Hills (named the Berkeley Hills in their paper) as overlapping the Mount Diablo subdivision and merging with the Mount Hamilton Range in the vicinity of Niles Canyon. The Mount Hamilton Range is least sharply delimited here, but the break formed by Niles Canyon

Fig. 3. Map of California; showing relationship of Diablo Range to other mountainous areas of California.

and Sunol Valley connects with the Santa Clara Valley on the west and Livermore Valley on the east and forms a farily well defined topographic, if not structural, boundary for this portion of the Mount Hamilton Range.

On the western side of the Mount Hamilton Range, the main western crest (Figs. 1 and 4) rises to an altitude of 3000 to 4000 feet, and consists, from north to south, of the following peaks and ridges: Oak Ridge (3280 feet), Mount Day Ridge (Mount Day, 3935 feet; Black Mountain, 3850 feet), Mount Hamilton (4209 feet), and Copernicus Peak[1] (4372 feet, the highest point in the Mount Hamilton Range and the third highest point in the Diablo Range as a whole), Mount Santa Isabella (4223 feet), Pyramid Rock (4014 feet), and Pine Ridge (3626 feet). South of Pine Ridge the western crest is not sharply defined. For the greater part of the length of the range the main western crest is paralleled by a subsidiary marginal ridge which

---

1 It forms the highest of the several peaks constituting the Mount Hamilton summit, and is usually considered as synonymous with Mount Hamilton.

rises sharply to an average height of 1000 feet above the Santa Clara Valley floor and is broken only by the trough formed by the entrance of Coyote Creek into the Santa Clara Valley. This marginal ridge is separated from the main western crest by Calaveras Valley in the north, Hall's Valley, San Felipe Valley, and Coyote Valley successively southward.

On the eastern side of the range, an eastern crest (the Red Mountains) parallels the western crest and reaches an altitude of over 3600 feet. This eastern crest extends almost the entire length of the range, but the highly distinctive topography of the Red Mountains is best developed in the summit area which marks the boundary line between Santa Clara and Stanislaus counties. Here the Red Mountains consist of a broken, rugged, mountain mass, with steep canyon walls of unstable talus which frequently lie at an angle of 35° or more (Fig. 5). To the east the Red Mountains decrease gradually in altitude to form a marginal foothill belt of rolling hills from five to fifteen miles wide, which disappears beneath the alluvial deposits of the San Joaquin Valley. In this foothill area differential erosion controls the drainage pattern, which consists of streams draining eastward into the San Joaquin Valley from the summit of the Red Mountains; this contrasts with the mainly north-south antecedent drainage pattern of the interior and western side of the range. The major eastern canyons thus formed are, from north to south, Corrall Hollow, Lone Tree Canyon, Hospital Canyon, Arroyo del Puerto, Salado Canyon, and Orestimba Canyon.

The interior of the Mount Hamilton Range, between western and eastern crests, consists of a mountainous area approximately five to ten or more miles wide, the topography of which is rough, but with no striking diversity of relief (Fig. 6). It represents a uniformly uplifted surface in which the original

Fig. 4. The northeast chaparral and forest covered slopes of Mount Hamilton from Packard Ridge; showing characteristic topography of the main western crest of the range.

Fig. 5. Red Mountains from the western slopes of Adobe Creek Canyon; showing the rugged, sparsely chaparral covered topography of this region. The steep talus deposits of Adobe Creek Canyon can be seen in the foreground.

Fig. 6. Sugarloaf Butte, and tributary to Arroyo Bayo; a typical view in the interior of the range.

erosional pattern has undergone little modification, the streams still flowing at grade in the old, undisturbed valleys (Willis, 1925). The ridges mainly parallel the longitudinal axis of the range, and reach a height of 2500 to 4000 feet, but their summit areas do not, in general, stand out as isolated peaks. Their slopes are only moderately steep, the streams, all of which drain into the Santa Clara Valley, occupy relatively wide channels which seldom lie below 2000 feet altitude, and in general the relief indicates a stage of early topographic maturity. Two fairly extensive valley areas occur in this interior portion of the range, the San Antonio Valley and the Santa Isabella Valley.

To summarize, the Mount Hamilton Range represents a distinct physiographic unit of the Diablo Range, its major features of relief best developed in a northwest-southeast direction paralleling the longitudinal axis of the range. The range can be rather easily divided into a western portion which arises abruptly from the Santa Clara Valley plains and consists of the flanking subsidiary ridge and the steep main western crest, an interior area, and an eastern portion which consists of the eastern crest (Red Mountains) and a foothill zone which drops gradually to the San Joaquin Valley plains. These major topographic features express the combined control of forces which elevated the range and the varying resistance to erosional forces of the rocks which compose it (Willis, 1925). The same factors have exerted a control upon plant distribution within the range, so that the major floristic changes coincide with the major topographic changes, and the floristic zones parallel the longitudinal axis of the range, dividing the flora into more or less distinctive western, interior, and eastern components as will be described.

GEOLOGIC HISTORY

Most of the surface area of the Mount Hamilton Range consists of rocks of the Franciscan series (Fig. 2). Because of the probable relationship of these rocks to the phytogeography of the Mount Hamilton Range, their rôle in the geologic history of the range will be described in some detail. The Franciscan rocks represent the oldest and most extensively exposed series of sedimentary rocks in the central and South Coast Ranges (Taliaferro, 1943). Their southern limit in the inner South Coast Ranges coincides approximately with the southern limit of the Diablo Range (Reed, 1933, fig. 16, p. 73). In the units of the Diablo Range south of the Mount Hamilton Range, they form a progressively narrower belt which is discontinuous at its southern extremity (Jenkins, 1938). In the Mount Hamilton Range they total 15,000 to 20,000 feet in thickness (Templeton, 1913), and compose the entire surface area of the range except on north, west, and east margins. At the northern margin of the range they dip beneath Cretaceous and Tertiary sediments. Northward in the Diablo Range they occur in small areas in the Contra Costa Hills and on the summit of Mount Diablo (Turner, 1891; Taft, 1935). In the inner North Coast Ranges they reappear in Lake County and adjoining areas of Napa, Sonoma, Colusa, and Glenn counties(vicinity of Clear Lake, Fig. 2), and northward they form a considerable portion of the surface layer of the North Coast Ranges. In the outer South Coast Ranges they occur in the

Santa Cruz Mountains and San Francisco peninsula; in the outer North Coast Ranges they are found from the Marin peninsula (Mt. Tamalpais region, Fig. 2) northward.

The Franciscan rocks had their origin as marine sediments laid down probably in the early Jurassic. During the Upper Jurassic they were highly folded and faulted and intruded with large quantities of igneous rocks. Consequently the Franciscan series is extremely complex and intensively altered, being composed of various minerals and many kinds of sedimentary, metamorphic, and igneous rocks mingled together. Sandstones, shales, cherts, and conglomerates are abundant among the sedimentary rocks (Davis, 1918a, b; Turner, 1891; Fairbanks, 1894). Large masses of jasper and variously related slates and schists characterize the metamorphic areas. The igneous rocks are mainly intrusive and ultrabasic or basic, common types being basalt, diabase, pyroxenite, gabbro, and peridotite (the latter now almost entirely altered to serpentine and associated ferromagnesian rocks). Throughout the Franciscan rocks of the Coast Ranges, serpentine, either of sedimentary or igneous origin, is frequent (Anderson and Pack, 1915; Anderson, 1905; Fairbanks, 1898; Turner, 1891), and from the San Carlos Range to Clear Lake in Lake County it is very abundant. North of Lake County the serpentine decreases in abundance along with a gradual change in other constituents of the Franciscan rocks.

Varying and sometimes contradictory accounts of the diastrophic history of the South Coast Ranges have been advanced (Willis, 1925; Clark, 1925, 1927, 1929, 1930, 1935; Reed, 1933). Their structure is highly complex. One of the most logical explanations is that the present structural units, including the Diablo Range of which the Mount Hamilton Range is a part, were brought into being by continued gentle compression along the continental margins upon a region of already diversified topography and structure (Taliaferro, 1943). Submerged troughs and elevated land masses had their start in the upper Cretaceous and Eocene, attained a continuous development in the Miocene, and reached their maximum in the strong diastrophisms of late Pliocene and early Pleistocene. By that time the present major structural and topographic features were developed, the Coast Ranges forming an archipelago system (Anderson, 1908; Lawson, 1893) in which the mountain ranges were insular land masses. The present exposure of the Franciscan rock series roughly coincides, at least in the Diablo Range, with these insular land masses. Thus the Mount Hamilton Range, as one of the structural units of the Diablo Range of the South Coast Ranges, can be considered an insular land mass which has been at least partly above water since the middle Miocene, at which time present-day climatic zones and a modern flora and fauna were beginning. The Mount Hamilton Range was partly or entirely surrounded by marine embayments until after the Pliocene-Pleistocene revolution, but its present exposure of Franciscan rocks is roughly correlated with that portion of the range which was above water during at least the later periods of coastal subsidence.

The western marginal ridge of the Mount Hamilton Range consists of

Cretaceous and Tertiary, mainly unaltered, marine sedimentary strata (shales, sandstones, conglomerates, etc.). It is in fault contact with the Franciscan rocks of the main part of the range (Sunol graben: Templeton, 1913; Vickery, 1925). The eastern marginal foothill strip consists of similar sediments (Anderson and Pack, 1915), and they also lie in fault relationship with the Franciscan rocks of the interior (Clark, 1927). In this area, however, the fault zone is not always direct or clear, and the strata are soft, so that differential erosion controls the topography to a far greater extent than on the western side of the range. The northern margin of the range likewise is composed of Cretaceous and Tertiary (and also Quaternary) sediments lying in fault relationship to the Franciscan rocks of the interior of the Mount Hamilton Range. To the south the Franciscan rocks continue across the Pacheco Pass into the San Carlos Range (Fig. 2).

Summarizing the geology of the Mount Hamilton Range, it is seen to be a land mass which has been above water probably since the early part of the Tertiary. It is composed mainly of altered Franciscan sediments and intruded basic igneous rocks, the Franciscan basement being overlain by and in fault contact with later, mainly unaltered marine sediments along the north, east, and west margins of the range. Although the Mount Hamilton Range has developed as a more or less structurally distinct land mass, its geologic history is similar to that of the other land masses of the Diablo Range, and more generally to all the South Coast Ranges. More of the total surface area of the Mount Hamilton Range belongs to the Franciscan series, however, than to any of the other units of the South Coast Ranges, and no larger expanse of Franciscan rocks occurs until the central portion of the North Coast Ranges is reached. These Franciscan rocks, due to their long history as exposed land masses, and to the distinctive soils they form, are believed to control, in part, the present distribution of plant species in the Mount Hamilton Range and in the central Coast Ranges as a whole, as discussed subsequently.

CLIMATE

Russell's classification (1926) of the climates of California is based on that of Köppen and gives a basis for determining the climatic provinces of the Mount Hamilton Range, and for analyzing, in generalized manner, the climate of the Mount Hamilton Range in relation to the Coast Ranges as a whole. According to Russell, the western and interior portions of the Mount Hamilton Range are characterized by a humid, mesothermal, cool-winter "Mediterranean" climate which is divisible into two types, a cool-summer "heather" type, in which the warmest month of the year averages below 71° F., and a hot-summer "olive" type, in which the warmest month averages above 71° F. From the summit of the main western crest of the range to its western base, the cool-summer type occurs, while in the interior of the range, the hot-summer type occurs. A Mediterranean climate of these two types is found throughout the greater part of the Coast Ranges. In the North Coast Ranges, the cool-summer "heather" type is predominant, and only two small areas of the hot-summer "olive" type occur (one in the vicinity of Clear Lake, Lake

County, the other in Mendocino County). It is significant to plant distribution that the small area of hot-summer climate in southern Lake County coincides with the area of Franciscan rocks found there (p. 9). The cool-summer "heather" type also is predominant in the northern part of the South Coast Ranges, but south of the Santa Cruz Mountains and the Mount Hamilton Range, it is restricted mainly to a coastal strip, while in the interior (Gavilan Range and Santa Lucia Mountains) the hot-summer "olive" type occurs.

The eastern side of the Mount Hamilton Range, from the summit of the eastern crest (Red Mountain) to the San Joaquin Valley, is characterized by an arid, hot, steppe climate. In the units of the Diablo Range south of the Mount Hamilton Range, this steppe climate predominates. It forms a marginal band in the foothills around the entire San Joaquin Valley, and gives way at the borders of the San Joaquin Valley to the arid, hot, desert (Mohave-type) climate of the latter. The steppe climate represents a climatic transition between the arid climate of the San Joaquin Valley and the humid climate of the major portion of the Coast Ranges.

Reviewing the above, it is seen that, in a general matter the Mount Hamilton Range can be divided into three longitudinal climatic zones, a western zone of cool-summer Mediteranean climate, an interior zone of hot-summer Mediterranean climate, and an eastern zone of arid steppe climate which is bordered by the arid desert climate of the San Joaquin Valley. These three climatic zones correspond, in general, to the three major topographic areas of the range, and in part the former are controlled by the latter (McAdie, 1903, 1914). Together these topographic areas and climatic zones are largely responsible for the longitudinal zonation of the flora within the Mount Hamilton Range.

The presence of three climatic zones in the Mount Hamilton Range implies a considerable climatic diversity. This becomes apparent when the major factors determining the climatic zonation are analyzed. Considering first rainfall, the higher prevailing westerly winds pass over the Mount Hamilton Range (and other parts of the Coast Ranges) entirely, dropping their moisture on the western slope of the Sierra Nevada, while the lower winds precipitate much of their moisture on the western side of the Santa Cruz Mountains before the Mount Hamilton Range is reached. Nonetheless, the steep western slopes of the Mount Hamilton Range act as a second effective bulwark in intercepting the lower prevailing westerly winds and in depriving them of much of their remaining moisture. Consequently the interior of the range receives considerably less rainfall than the western slopes. The eastern crest of the range catches some of the moisture left in the lower winds, draining it into the interior of the range, so that the eastern foothill area receives even less precipitation than the interior area. Relative humidity, temperature, and atmospheric pressure are affected as well, although to a lesser extent than rainfall, and as a result the climate of the eastern side of the range is markedly continental in comparison to the maritime climate on the western side of the range.

On the western crest, the summit of Mount Hamilton (including Copernicus Peak, highest point in the range) has a maximum rainfall for the range. Rain occurs every month of the year on Mount Hamilton, but the summer rains are light and infrequent. Over half the annual rainfall occurs between December and March. During the winter months the precipitation is frequently in the form of snow. The average seasonal precipitation for Mount Hamilton is 32.28 inches.

No rainfall data are available for the interior area of the Mount Hamilton Range. Some idea of the average seasonal precipitation can be obtained, however, by considering the data from other areas in the South Coast Ranges which lie in the same hot-summer climatic zone. It is probably safe to assume that the rainfall in the interior of the Mount Hamilton Range is about 15 inches, or less than half that of the western crest. July and August are almost rainless in the interior. The streams are mainly intermittent, and carry running surface water only in the winter, spring, and early summer months.

The average annual precipitation for the eastern side of the Diablo Range (which would include the eastern side of the Mount Hamilton Range as well) is only 9 inches according to Anderson and Pack (1915). The eastern side of the Diablo Range receives little or no more average rainfall than the upper end of the San Joaquin Valley at the base of the Tehachapi Mountains.

Summer fogs are frequent on the western slopes of the Mount Hamilton Range, and are important in the development of the strongly maritime climate which characterizes this side of the range (Byers, 1930). The Golden Gate and San Francisco Bay (Fig. 2) furnish the main path of entrance for the fogs, from whence they are carried southward down the San Francisco Bay to spread out in the Santa Clara Valley and on its flanking slopes. Frequently, however, the outer Coast Ranges from San Mateo County to Mount Tamalpais are low enough to allow the fogs to drift in over their summits. Chittenden Pass, between Monterey Bay and the southern end of the Santa Clara Valley, also acts as a path of entrance for fogs. The fogs are often high enough to reach the summit areas of the western crest of the Mount Hamilton Range. Occasionally they may spread via Niles Canyon into Livermore Valley at the northern end of the Mount Hamilton Range.

Temperature varies considerably between western, interior, and eastern portions of the range. It is, of course, hottest, and the diurnal and seasonal extremes are greatest on the eastern side, but no accurate data are available. The mean annual temperature of Mount Hamilton is 52.7° F., the mean temperature of the coldest month (January) is 39.7° F., and the mean temperature of the warmest month (July) is 69.4° F. Temperature appears to be of less importance than rainfall as a climatic factor in the Mount Hamilton Range. Russell (1926) corroborates this by stating that the steppe climate of the San Joaquin Valley and inner portion of the South Coast Ranges (including the eastern side of the Mount Hamilton Range) falls into this subdivision not on the basis of temperature, but primarily because of "rain shadow" conditions caused by areas of higher altitude toward the west.

SOILS

The surface area of the Mount Hamilton Range, as already stated, is almost entirely composed of rocks of the Franciscan series. These rocks were highly fractured as a result of the diastrophic forces exerted upon them during the formation of the Mount Hamilton Range. As a consequence, the Franciscan rocks, although predominantly metamorphic and igneous in composition, are easily fragmented by weathering. On the ridges of the Mount Hamilton Range, Franciscan bed rock is frequently exposed, but often all bed rock is covered with more or less disintegrated mantle rock. The slopes of the ridges tend to form unstable rock slides, the size of the constituent rock fragments depending largely upon the degree of fracturing of the bed rock. Such talus deposits are a characteristic feature of the topography throughout the interior of the range, but they reach their highest development in the Red Mountains (Figs. 7 and 8). Here the serpentine which composes the area is very friable. The arid climate with rain concentrated during short periods, winds, gravity,

Fig. 7. A precipitous, unstable, almost barren talus deposit in Adobe Creek Canyon of the Red Mountains.

Fig. 8. Adobe Creek Canyon of the Red Mountains; showing again its precipitous slopes and unstable talus deposits.

and sparseness of the vegetational cover, all combine to weather and erode the serpentine into steep, unstable, porous rock slides of relatively small fragments (mainly rubble or gravel) which often extend 2000 feet or more from ridge top to canyon floor. On talus deposits of this type, a true soil scarcely can be said to exist. It is a skeletal soil, predominantly mineral in composition, and chiefly conditioned by the character of the underlying rock rather than by climate or vegetational cover. In these areas plant growth is usually highly limited, both in numbers of species and individuals. A few species show adaptations to the unstable substratum. *Psoralea californica*, for example, has only a low tuft of leaves and stems above ground, but its stout, deeply seated root may extend twelve inches or more underground.

Under the influence of an even more arid and hot climate, the soft, unaltered sediments which occur east of the Red Mountains have weathered in essentially the same manner as the Franciscan rocks of the interior and eastern crest of the range, producing porous soils which are deficient in organic matter. In general, however, the soils from these sedimentary rocks are much richer in lime than the soils formed from the Franciscan rocks, and their vegetational cover is not so sparse.

Throughout the interior and eastern side of the range, well developed soils are found only on the more gentle slopes (mostly grassland or savannah areas), in those portions of the canyon floors not occupied by the wide gravelly flood channels of the streams, and in the occasional valley areas. On the western slope of the range and the upper reaches of the western crest, in both sedimentary and Franciscan rocks, well developed soils are much more prevalent, and a soil climax is more generally approached. In this area, soil formation is chiefly conditioned by the complementary effects of a more maritime climate and a more extensive vegetational cover rather than by the nature of the underlying rock. In the forested areas of the sheltered ravines and upper slopes of the western crest, the soil is rich in humus.

As indicated above, climate, vegetational cover, and the physical nature of the rocks involved, have played interacting parts in the production of soil in the Mount Hamilton Range. Conversely, the degree to which development of the soil has proceeded in various portions of the range determines in part the relative abundance and nature of the vegetational cover of the range.

Another related factor which exerts a very considerable influence upon plant distribution within the range is the chemical nature of certain of the soils. It has been pointed out that ultrabasic or basic, intrusive, igneous rocks form a large and important element in the Franciscan series. Particularly abundant among these igneous intrusives is the group of ferromagnesian minerals of which peridotite was originally the most important constituent. The rocks of this group are classed as basic to ultrabasic (i.e., having a low silica content, little or no free quartz, and usually with large amounts of iron), and the peridotite, being very subject to alteration, has been transformed almost entirely into serpentine. The physical and chemical processes of weathering ultimately break down the serpentine into brown, ferromagnesian soils which represent a mixture of carbonates and silicates (Pirsson, 1926). The soils so formed lack potassium and calcium, and are high in magnesium content.

Such serpentine soils are often exceedingly infertile. According to Gordon and Lipman (1926), the infertility of these soils is due, not to a specific toxicity induced by their high magnesium content, but to four causes: 1) they have a very low concentration of available ions which are important to plant growth, 2) they have a low nitrate content, 3) they have a high pH value (averaging 8.1 on the samples tested), and 4) they are poor in potassium and phosphates. Of these, Gordon and Lipman believe the high pH to be the most significant factor in rendering serpentine soils infertile.

In the Mount Hamilton Range, the Red Mountains form the major serpentine area. They are composed of iron-impregnated serpentine rocks with occasional outcroppings of magnesite, manganese, and silicious rocks (Calif. St. Min. Bur. reports, 1893, 1915). The iron imparts a red color to the rocks, from which the mountains derive their name. The infertility of the soils in the Red Mountains is strikingly exemplified by their very sparse vegetational cover. There are also occasional bands of serpentine along the western crest of the Mount Hamilton Range. On the Mount Day Ridge, the serpentine rocks are compact, waxy, and greenish, very different in appearance from the more heavily iron impregnated serpentine of the Red Mountains.

Throughout the South Coast Ranges serpentine is a frequent constituent of the Franciscan rocks. The Mount Diablo summit includes a narrow band of serpentine and associated igneous rocks. Serpentine is also common in the southern portion of the Franciscan series of the North Coast Ranges.

Although serpentine frequently inhibits plant growth, its peculiar qualities may cause it to favor the growth of certain species which require an excess of magnesium ions in relation to calcium ions. Serpentinous species usually have a highly developed root system and poorly developed surface parts (Braun Blanquet, 1932). A limited number of plant species appear to be obligate to areas of serpentine rocks, and in these cases plant distribution is primarily a matter of local, edaphic control. This is borne out in the distribution of certain species in the Mount Hamilton Range, such as *Allium Parryi, A. fimbriatum, Fritillaria folcata, Salix Breweri, Quercus durata, Streptanthus Breweri, Linum Clevelandii, Clarkia Breweri, Garrya Congdoni, Emmenanthe penduliflora* var. *rosea, Collomia diversifolia,* and *Cirsium campylon.*

As pointed out by Gordon and Lipman (1926), the characteristics of a serpentine-derived soil also occur in other magnesium-high soils. As a whole, magnesium rocks are not good soil formers. Although serpentine is the most abundant magnesian rock of the Franciscan series, other magnesian minerals and rocks are frequent (magnesite; olivine, a mineral constituent of basalt and peridotite; pyroxene, a common mineral in pyroxenite and metamorphic schists; etc.), and the soils they form exert a control upon plant distribution similar to that of serpentine.

Other igneous rocks of the Franciscan series, and likewise some of the metamorphics, are poor soil formers. Schists and gabbro, unless associated with accessory minerals, give infertile soils. Jasper, a common mineral among Franciscan rocks in the South Coast Ranges, is mainly silicon dioxide, and gives an inert soil.

The chemical nature of the soils appears to influence plant distribution in the Mount Hamilton Range throughout the surface area of the Franciscan rocks, although its effect is most strongly exhibited in the serpentine areas. Other environmental factors being equal, plant distribution is affected similarly in other portions of the Coast Ranges where these rocks appear in abundance. This accounts, in part, for the limited and discontinuous distribution of certain inner Coast Range species to be discussed. It is noteworthy that north of Lake County the Franciscan rocks undergo a gradual change, the serpentine and jasper constituents becoming less abundant, and hornblende, feldspar, and mica-like schists and slates becoming more frequent. The latter group forms good soils, and this is probably of significance in the northward restriction of certain of the central Coast Range endemic species.

In conclusion it is important to note that the influence which the chemical nature of the soils, particularly serpentine, has upon plant growth, species evolution, and species distribution, forms the basis of one of the most salient and interesting of the many baffling distributional problems involved in the California flora. Any satisfactory solution of this problem must await the combined efforts of plant physiologist and systematic botanist.

INFLUENCE OF MAN

In the Mount Hamilton Range the biotic influence expresses itself today not so much as a factor in the establishment and maintenance of the native plant species as in their restriction and extermination. The erection of the Lick Observatory upon the summit of Mount Hamilton in 1876 has resulted in year-round human occupancy of this area. The establishment of a permanent colony here, and the subsequent influx of tourists and visitors has led to a marked local alteration of natural conditions.

Observatory residents report that some of the native species which were frequent on the summit area of Mount Hamilton twenty or more years ago are now uncommon or very rare. Among these species are *Clarkia Breweri, Arabis Breweri,* and *Calochortus venustus.* Their gradual restriction in this region no doubt is due not only to the disturbed conditions brought about by human occupancy and the resulting competition with introduced weeds, but also to the rapid increase of deer following the establishment of a state game preserve on the upper slopes of the mountain in recent years.

The lower western slopes of the range have been intensively cultivated for many years. Here the original flora has almost disappeared, both as a result of cultivation and of the competition offered by the invasion of alien species, for it is in this area that most of the aliens are definitely established. As an example, *Avena fatua* is abundant here, while in other portions of the range it is merely adventive.

The Calaveras Reservoir has flooded the Calaveras Valley and the lower portion of the Arroyo Hondo, exterminating the flora of this region (and exterminating *Acanthomintha lanceolata* Curran at its type locality).

Except in the areas already mentioned, the native vegetation of the Mount Hamilton Range as a whole has been relatively little disturbed by the influences of man, domesticated animals, and the concomitant changes in the native fauna. The interior of the range and particularly the eastern side are very sparsely settled, although these areas have been long subject to cattle grazing. Most of these areas are still open range, and only isolated patches of ground are under cultivation. On the eastern side of the range, sheep have been grazed since a time prior to Brewer's botanical explorations in 1862. Part of the paucity of the vegetational cover here may be a result of overgrazing. The only significant modification in the native flora of the interior and eastern side of the range, however, is the establishment in the grassland areas of a number of alien grasses, such as *Bromus rubens, B. mollis, Hordeum murinum, Lolium multiflorum, Koeleria cristata,* and *Avena barbata,* and the extensive establishment of *Erodium cicutarium.*

Analysis of the Vegetation

Plant Communities

The California Forest and Range Experiment Station of the Division of Forestry, United States Department of Agriculture, has completed the following vegetation type maps (not yet published) of the Mount Hamilton Range: Carbona, Pleasanton, Tesla, Mount Hamilton, Morgan Hills, and Orestimba quadrangles. Maps are in progress for the other portions of the range. Due to this work and because the emphasis of this paper is floristic rather than vegetational, only a brief description of the more important plant communities and their occurrence will be given.

In the vicinity of Mount Hamilton, the western slopes of the range are mainly grassland. Farther south, the chaparral of the interior extends over the summit of the western crest and covers the western slopes scantily or thickly. The grassland on the western side of the range often supports a scattered growth of oaks, while patches of a sagebrush community (mainly *Artemisia californica*) are frequent on the dryer slopes. In the more mesophytic areas (canyons and north-facing slopes), the grassland gives way to brush and woodland (deciduous thicket scrub and broad sclerophyll forest of Cooper, 1922). Species typical of the brush are *Ribes malvaceum, Rosa californica, Holodiscus discolor, Amelanchier alnifolia, Osmaronia cerasiformis, Toxicodendron diversiloba, Symphoricarpos albus, Sambucus coerulea,* etc. Species typical of the woodland are *Quercus agrifolia, Q. chrysolepis, Q. Kelloggii, Q. Wislizenii, Umbellularia californica, Acer macrophylum,* and *Rhamnus californica* subsp. *tomentella.* The higher summit areas of the western crest support a mixed coniferous-broad sclerophyll forest. The typical broad sclerophyll elements in this forest are the same as the woodland species listed above, while *Pinus Coulteri* is the only important conifer. On the northerly slopes of the western crest or in the canyons on the western slope, however, local conditions support occasional pockets of *Pinus ponderosa,* a species usually found at somewhat higher altitudes. There is also the infrequent occurrence, in similar pockets, of typical outer Coast Range species such as *Torreya californica* and *Arbutus Menziesii.*

In the interior of the Mount Hamilton Range, the climax chaparral association (sensu Cooper, 1922[2]) is the dominant community. The most important species in this highly xerophytic community, in the Mount Hamilton Range as elsewhere in California, is the chamise, *Adenostoma fasciculatum.* It occupies large areas in the interior of the range as a pure or almost pure dominant. *Ceanothus cuneatus* is the second most important constituent of the climax chaparral association, and other representative species are *Quercus dumosa, Cercocarpus betuloides, Photinia arbutifolia, Prunus ilicifolia, Ceanothus leucodermis, Rhamnus crocea* subsp. *ilicifolia, Garrya Fremontii, Arctostaphylos glauca, A. glandulosa* var. *Campbellae, Eriodictyon californicum,* etc.

---

2 Cooper includes two associations under chaparral (broad sclerophyll scrub) formation—climax chaparral association and coniferous forest chaparral association.

Extensive areas of grassland or savannah (grassland supporting a thin growth of *Pinus Sabiniana, Juniperus californicus,* or various oak species, mainly *Quercus Douglasii* and *Q. lobata*) also occur in the interior of the range, particularly in the valleys or on their adjacent slopes (Fig. 6). North-facing slopes of ridges support occasional patches of brush or woodland, although these are composed of less mesophytic species than similar areas on the western slopes of the range. Typical species are *Berberis dictyota, Ribes quercetorum, Ribes malvaceum, Prunus emarginata, Forestiera neomexicana,* and *Lonicera subspicata* var. *Johnstoni.*

According to Cooper (1922), chaparral formerly controlled much of the area now occupied by grassland in the Coast Ranges. As an illustration he describes (p. 78) the "patchy remnants of chaparral mainly *Adenostoma*" found on the grassy hills of the western slopes of the Mount Hamilton Range above Coyote Creek, while toward the interior of the range the "numerous ridges are covered with an irregular mosaic of chaparral and grassland," deeper penetration finally bringing one to a central region of solid chaparral. Wherever mature individuals of *Adenostoma fasciculatum* are found in grassland, either isolated or in patches, Cooper believes that chaparral was formerly dominant. Following Cooper, this would indicate that most of the present areas of grassland and savannah on the western side and interior of the range may at one time have been occupied by climax chaparral, fire having been the main factor involved in its destruction.

On the eastern crest of the Red Mountains area, a thin cover of chaparral occurs on the more stable talus deposits. Here again the chaparral is predominantly or exclusively *Adenostoma fasciculatum,* but serpentinophilous species are frequently intermingled, the most characteristic being *Quercus durata* and *Garrya Congdoni.* As already mentioned, the more unstable talus deposits, which give the Red Mountains their distinctive topography, are very barren. They are almost or entirely devoid of shrubs, and the herbaceous cover is widely spaced. The latter consists of annual or perennial species which are adjusted to an unstable, highly xeric, and ecologically distinctive substratum, of which the following are typical: *Allium Parryi, A. fimbriatum, Streptanthus Breweri, Psoralea californica, Linum Clevelandii, Clarkia Breweri,* and *Campanula exigua.* Grassland is poorly represented in this region.

In the eastern foothill area, chaparral and savannah or grassland are frequent, the latter becoming predominant as the eastern margin of the range is approached. North-facing or canyon slopes seldom support brush or woodland thickets as in the interior of the range. A sagebrush community, consisting of such species as *Artemisia californica,* which is usually dominant, *Eriogonum fasciculatum* var. *foliolosum, Malvastrum Fremontii* var. *cercophorum, Salvia mellifera,* and *Mimulus aurantiacus,* forms occasional islands in the grassland.

It is likely that the grassland of the extreme eastern side of the range was not formerly dominated by chaparral as postulated by Cooper (1922) for the interior and western sides of the range. This area closely approaches the climatic conditions of the San Joaquin Valley, where grassland is considered to be the climax. Likewise, there are many species of desert affinities in this eastern foothill zone, such as *Prosopis chilensis, Oenothera deltoides* var. *cognata, Nicotiana glauca, Amsinckia vernicosa, Plagiobothrys arizonicus, Salvia carduacea, Malacothrix Coulteri*, etc. Their presence indicates a close environmental relationship with the desert-like ridges which surround the southern end of the San Joaquin Valley where chaparral, according to Cooper, probably never controlled the present grassland areas. Even toward the interior of the range on this eastern side, a few typically desert species are found in the grassland, for example, *Eriophyllum Wallacei* and *Lepidospartum squamatum*.

A riparian community occurs along the perennial streams throughout the range, but is most highly developed on the western side, as for example, along Arroyo Hondo and its tributaries, Smith Creek and Santa Isabella Creek. In these areas, thick canopies of *Alnus rhombifolia, Acer macrophyllum, Arbutus Menziesii, Platanus racemosa*, and *Umbellularia californica* provide shaded niches for such mesophytic herbaceous species as *Adiantum Jordani, Cystopteris fragilis, Disporum Hookeri, Smilacina amplexicaulis, S. sessilifolia, Epipactis gigantea, Aquilegia Tracyi, Mimulus nasutus*, and *M. cardinalis*. Where the streams are intermittent, as they mainly are in the interior and on the eastern side of the range, the riparian community is much more limited and much less mesophytic, and scattered trees of *Salix laevigata, S. lasiolepis, Populus Fremontii*, or *Platanus racemosa*, together with certain species of *Carex* and *Juncus*, may represent the only definitely stream-side species.

If, in summary, a comprehensive picture of the vegetation is attempted,

Fig. 9. North end of Adobe Valley at eastern edge of Red Mountains; typical savannah in foreground, dense chaparral slope in background.

the major communities appear to be grassland and chaparral. On the western slope grassland usually predominates; in the interior of the range grassland or savannah are abundant, but chaparral covers large areas; on the eastern crest (Red Mountains) chaparral is the most extensive community, while in the eastern foothill belt grassland usually predominates. Thus, from the vegetational viewpoint, no one portion of the range is markedly diverse from any other portion; the several communities are repeated in various areas.

When the floristic content of the communities is considered, however, marked changes are seen to occur as one crosses the narrow axis of the range (i.e., latitudinally, or from west to east), so that four longitudinal floristic zones are evident: 1) a western zone which includes the western slope and crest of the range, 2) an interior zone, that area which lies between western and eastern crests, 3) the eastern crest (Red Mountains), 4) an eastern foothill zone. In part these floristic zones are controlled by topography and climate, the western zone containing the most mesophytic species, the eastern foothill zone containing the most xerophytic (here occur most of the Great Basin and desert derivatives discussed later). In the interior and especially in the Red Mountains, however, the influence of the soils and the long history of these regions as exposed and isolated land masses are both significant. It is here that most of the central Coast Range endemic species occur.

SEASONAL PERIODICITY

There is a seasonal periodicity involved in the development of the vegetation, with two well marked seasons of plant growth, vernal and aestival, each characterized by a different flora. The vernal flora is composed of many species of ephemeral annuals, the individual plants of which occur in great numbers and cover large areas of grassland throughout the range. The vernal flora is particularly well developed in valley areas in the interior of the range, such as Santa Isabella Valley and San Antonio Valley. It is also well developed in the eastern foothill zone. The life span of the vernal annuals is short, and is adapted to the period of moderate temperature which follows the late winter and early spring rains and precedes the aridity and high temperatures of summer. Flowering usually occurs in late March or early April. Such a vernal flora is typical of large areas in the grasslands of California. Some of the characteristic species in the Mount Hamilton Range are *Ranunculus californica, Athysanus pusillus, Eschscholtzia californica, Lupinus bicolor, Gilia tricolor, Nemophila Menziesii, Orthocarpus purpurascens, O. densiflorus, Achyrachaena mollis, Layia platyglossa, Coreopsis calliopsidea, Monolopia major*, etc.

The aestival flora is represented by a relatively few species of annuals, herbaceous perennials, or suffrutescent plants which are adapted to a highly xeric environment. The number of individuals produced, however, and the area covered, are large. Again the grasslands of the range are the major areas occupied. The aestival flora reaches its best development in the interior and on the eastern side of the range. Flowering usually occurs in late August and September during the period of highest temperatures and little or no rainfall.

Like the vernal flora, this aestival flora is characteristic of the grasslands of California. Typical species in the Mount Hamilton Range are *Eriogonum vimineum, E. virgatum, E. Wrightii, Eremocarpus setigerous, Grindelia camporum,* and various species of *Hemizona, Calycadenia,* and *Madia.*

Although vernal and aestival seasons of plant growth are highly differentiated, by no means all species are adapted to one or the other of these seasonal patterns. Many species bloom mainly in late spring and early summer (May and June), although these are not so well represented as the vernal or aestival species, either from the standpoint of species involved, numbers of individuals, or the areas occupied. From a taxonomic and geographic viewpoint, however, many of the most distinctive species bloom in the late spring. These species are best developed in the interior and on the eastern side, and characteristic among them are: *Allium falcifolium, Eriogonum saxatile, Streptanthus callistus, Sedella pentandra, Clarkia Breweri, Oenothera decorticans* var. *typica, Sanicula saxatilis, Phacelia Breweri, Salvia Columbariae, Acanthomintha lanceolata, Castilleja Roseana, Malacothrix floccifera, Cirsium campylon,* etc.

Geographical Analysis of the Species

Although the vascular species of the Mount Hamilton Range comprise a relatively distinctive assemblage, over half being endemic to California, the flora is not a homogeneous one. Like all other "floras" it is composed of diverse floristic elements (groups of species), each homogeneous in that it has had a common center of origin and dispersal from which the component species migrated. These elements arose in response to certain physical and biological factors or interacting combinations of these factors, and the species within each element have a similar migrational history and a basically similar pattern of distribution. In reconstructing the geographical history of a flora, it should be theoretically possible to determine the composition, origin, and migration of each of its floristic elements. A more or less arbitrary disposition of many species is necessary, however, for paleontological evidence may be meager or lacking, many morphological and taxonomic affinities are obscure, genetic relationships are infrequently and incompletely known, distributional knowledge is seldom complete, and centers of origin and dispersal together with the areas of greatest differentiation are difficult to trace. Therefore the floristic elements, as recognized, do not represent completely homogeneous groups of species, and their true composition is only partly realized. With these points in mind, division of the Mount Hamilton Range flora into its component floristic elements may be attempted. Two groups of species of dubious history, and three floristic elements are considered in the following discussion. Because of the limits of knowledge concerning them, the elements, as presented, vary greatly in their homogeneity. Origin and migration of the species in each element are considered only where they aid directly in analyzing the flora of the Mount Hamilton Range.

COSMOPOLITAN SPECIES

A group of cosmopolitan species occurs in the Mount Hamilton Range, these species having a very broad climatic tolerance and being more or less world wide in distribution, at least as to the extratropical zones of the northern hemisphere. The following list indicates the types of cosmopolitan species which occur in the Mount Hamilton Range, but is not intended to be a complete enumeration. Many of the species are of aquatic or semi-aquatic habitats. The origin and migration of these widely-distributed species is exceedingly difficult to trace, and the products of several floristic elements are probably represented. Species of this type are of little value in determining the history of a local flora such as that of the Mount Hamilton Range.

Cystopteris fragilis	Juncus bufonius
Equisetum arvensis	Montia fontana
Zannichellia palustris	Spergularia saligna
Alisma Plantago	Ranunculus trichophyllus
Phragmites communis	Radicula Nasturtium-aquaticum
Hordeum nodosum	Arabis glabra
Eleocharis mammilata	Capsella procumbens
Eleocharis acicularis	Callitriche palustris
Lemna minor	Galium Aparine
Juncus sphaerocarpus	Artemisia Dracunculus
Juncus balticus	

INTRODUCED SPECIES

Nine per cent of the Mount Hamilton Range flora consists of introduced or alien species. This includes both definitely established and merely adventive species. Their relative influence in competition with the native vegetation was discussed under the influence of man. These species may be European, Asiatic, or Australian in origin, and they compose a homogeneous group only in the sense that their history of migration and establishment in new areas is coincident with or follows that of man. They are not a part of the natural plant life of the Mount Hamilton Range. The known introduced species are:

Bromus rubens	Amaranthus blitoides
Bromus rigidus	Glinus lotoides
Bromus mollis	Cerastium viscosum
Bromus arenarius	Stellaria media
Festuca dertonensis	Sagina apetala var. barbata
Poa annua	Herniaria cinerea
Poa pratensis	Silene gallica
Lamarckia aurea	Brassica campestris
Hordeum murinum	Brassica arvensis
Hordeum gussoneanum	Brassica incana
Lolium multiflorum	Capsella Bursa-pastoris
Avena fatua	*Prunus cerasifera
Avena barbata	*Pyrus Malus
Agrostis verticillata	Medicago lupulina
Polypogon lutosus	Medicago hispida
Polypogon monspeliensis	Medicago apiculata
Gastridium ventricosum	Melilotus alba
Urtica urens	Vicia sativa
Rumex crispus	Geranium dissectum
Rumex conglomeratus	Erodium Botrys
Rumex acetosella	Erodium moschatum
Chenopodium album	Erodium cicutarium
Salsola Kali var. tenuifolia	Conium maculatum

* = Sporadic escapes from cultivation, and not established.

Anagallis arvensis
Marrubium vulgare
Lamium amplexicaule
Nicotiana glauca
Verbascum thapsus
Plantago major
Dipsacus Fullonum
Hypochaeris glabra

Lactuca saligna
Sonchus asper
Xanthium spinosum
Matricaria matricarioides
Senecio vulgaris
Cirsium lanceolatum
Centaurea melitensis

NORTHERN FLORISTIC ELEMENT

At least fifteen per cent, probably a much larger proportion, of the Mount Hamilton Range species belong to the so-called "northern" floristic element of the California flora. These are species presumed to have had an origin in the north; or some may represent warm-temperate descendants of species which had a southern origin and spread north in the Eocene, later reinvading southern areas. During the middle of the Tertiary they formed a widespread, uniform flora over much of the northern portion of the northern hemisphere, this flora being known as the Miocene redwood forest of western North America. Southward migration and climatic segregation of these northern species began in the late Miocene and continued through progressive climatic changes to the end of the Tertiary, giving way to localized segregation in the Quaternary. The total composition of the northern floristic element in California is by no means fully known. Paleobotanical evidence has established the migrational history of the Miocene redwood forest, however, (Chaney, 1926) and a number of Mount Hamilton Range species are present-day equivalents of species which belonged to this redwood flora. In addition, many Mount Hamilton Range species not known to have equivalents in the redwood flora, including many species now restricted to California, are believed, on the basis of other criteria, to belong to the northern floristic element. Representative examples are:

Torreya californica
Scribneria Bolanderi
Trillium sessile var. giganteum
Disporum Hookeri
Habenaria unalaschensis
Alnus rhombifolia
*Polygonum Parryi
Claytonia gypsophiloides
Arenaria pusilla
Umbellularia californica
Berberis pinnata
*Paeonia Brownii

*Isopyrum stipitatum
Thalictrum polycarpum
*Platysperum scapigerum
Ribes sanguineum var. glutinosum
Prunus emarginata
Rosa gymnocarpa
Amelanchier alnifolia var. subintegra
Trifolium cyathiferum
Acer macrophyllum
Arbutus Menziesii
Crepis occidentalis subsp. pumila
*Crepis monticola
*Crocidium multicaule

* = Known southern limit in the Mount Hamilton Range.

Many of the northern species which occur in California have a relatively wide range of climatic tolerance, and are found throughout most of the state. Others, however, are limited mainly or entirely to the cool, humid climate of the North Coast Ranges. Some of these more restricted northern species occur in the outer South Coast Ranges, and a few also occur in portions of the warmer, more arid inner South Coast Ranges. One small although distinctive group of species from the preceding list is well represented in the North

Coast Ranges, although it occurs in the South Coast Ranges only in the Mount Hamilton Range. These species can be classed as "northern" with considerable assurance, although there is no paleobotanical evidence available that they belonged to the redwood flora. They are: *Scribneria Bolanderi, Polygonum Parryi, Arenaria pusilla, Paeonia Brownii, Platyspermum scapigerum, Trifolium cyathiferum, Crepis occidentalis* subsp. *pumila,* and *Crocidium multicaule. Eriogonum hirtiflorum, Isopyrum stipitatum, Viola Sheltonii,*[3] and *Crepis monticola,* although not found north of southern Oregon, appear to belong to this group as well. Stebbins (1938) and Babcock and Stebbins (1939), treating *Paeonia Brownii* and the two *Crepis* species respectively, consider these species to be reliquial in the Mount Hamilton Range, climate being the controlling factor in their persistence. It is possible that investigations may yield more or less similar interpretations for all the above species.

As to *Paeonia Brownii,* Stebbins believes it to represent a species which shows definite conservatism in the southern part of its range, where it is dying out due to the progressive advance of a warmer climate. Thus the Mount Hamilton Range locality represents the only known and presumably the last outpost of *P. Brownii* in the South Coast Ranges. As to the two *Crepis* species, Babcock and Stebbins consider the polyploid forms of these two heteroploid species to have a wide range of tolerance. The Mount Hamilton Range is a distant outpost for both, where they exist as single, static, apomictic forms, or biotypes, highly isolated from the centers of dispersal of the diploid sexual forms where new biotypes are being produced constantly. These two Mount Hamilton Range apomicts have survived on the distributional fringes of each species because their climatic tolerance was such as to permit them to survive under changing environmental conditions which eliminated the host of other biotypes. They are genetically depleted types, able to persist at their present outposts only so long as the environment remains favorable to their climatic tolerance (Stebbins, 1942).

In connection with the reliquial nature of *Paeonia Brownii* and *Crepis occidentalis* subsp. *pumila* in the Mount Hamilton Range, it is significant that both these species are represented in southern California by closely related but genetically distinct units. *Paeonia Brownii* is replaced there by *P. californica,* a species adapted to and advancing with the warmer, more arid climate of that area (Stebbins, 1938). *Crepis occidentalis* subsp. *pumila* is represented in southern California by an apomictic form related to Sierra Nevada rather than Coast Range biotypes, and thus with a different migrational history and range of tolerance than the Mount Hamilton plants (Babcock and Stebbins, 1939). Of the northern species being discussed, *Eriogonum hirtiflorum* is replaced in southern California by the closely related *E. inerme* which comes north as far as the Mount Hamilton Range (and Mount Diablo?). *Arenaria pusilla, Polygonum Parryi,* and *Trifolium cyathiferum* reappear in southern California, although highly localized. Presumably they are reliquial there and in the Mount Hamilton Range. On the other hand, cytogenetic study may indicate that, for these species as for the *Eriogonum, Paeonia,* and *Crepis* species, the

---

3 Also found in one locality on Mount Diablo.

southern California and Mount Hamilton Range plants represent different genetical strains adapted to different climatic conditions.

Sonoran Floristic Element

Certain species, here designated as the Sonoran floristic element, are also known in the California literature as the Sierra Madrean element (Axelrod, 1939), the Californian element, the southern or austral element, or the southwestern element, each of these names representing slightly different points of view as to origin and history of the species involved. As with the northern element, complete knowledge of the composition of the Sonoran floristic element in the California flora is lacking. Less paleobotanical evidence is available upon which to establish the origin and migrational history of this element than for the northern element, although recent papers by Axelrod (1938, 1939) concern it and its segregational products. According to Axelrod, this element arose in the Sierra Madre area of northern Mexico during the Oligocene, and migrated northward along arid, upland routes in the Miocene and later Tertiary in response to the progressive aridity of the climate in northern Mexico, southwestern United States, and southeastern California. It was more continental in climatic requirements than the northern element, and consequently was quite widely separated from the latter during the Miocene. Invasion of the lowland and northern areas was not effected to any extent until the dryer climate of the Pliocene made this possible. The modern representatives of this flora have their present centers of distribution largely in northern Mexico and the southwestern United States, but extend north over the Columbia Plateau, east to Oklahoma, and west to California. The Great Basin and desert species which enter California are probably segregates of the Sonoran floristic element, as well as many species now restricted to California. The Californian derivatives are concentrated in desert and cismontane areas of southern California, but many extend north to central or even northern California, or are now limited to areas north of southern California. The Californian derivatives constitute the "Californian element" of Axelrod (1939). He regards them as a segregation product of the Sonoran floristic element, which became predominant in southeastern California in the late Pliocene following climatic changes that involved a shift in seasonal rainfall from summer to winter months.

Following is a representative list of Mount Hamilton Range species believed to have arisen from the Sonoran floristic element:

Selaginella Bigelovii	Allium Parryi
*Pinus Coulteri	Allium peninsulare var. crispum
*Juniperus californicus	Muilla serotina
Stipa pulchra	Calochortus invenustus
Stipa lepida	Calochortus clavatus
Puccinellia simplex	Quercus Douglasii
Allium fimbriatum	*Quercus dumosa
Allium lacunosum	*Quercus chrysolepis

* = Species with known fossil records or for which equivalent species are known in the fossil record.

*Salix lasiolepis
*Populus Fremontii
Chorizanthe perfoliata
Chorizanthe polygonoides
Chorizanthe Clevelandii
Chorizanthe uniaristata
Eriogonum nudum var. auriculatum
Eriogonum virgatum
Eriogonum saxatile
Eriogonum fasciculatum var. foliolosum
Eriogonum Wrightii
Atriplex Serenana
Calyptridium monandrum
Calyptridium Parryi
Umbellularia californica[4]
Delphinium Parryi
Papaver heterophyllum
Tropidiocarpum gracile
Streptanthus Coulteri var. Lemmonii
Lithophragma Cymbalaria
Ribes quercetorum
Ribes aureum var. gracillimum
Ribes malvaceum
Rives amarum
Ribes speciosum
*Platanus racemosa
*Photinia arbutifolia
*Cercocarpus betuloides
*Holodiscus discolor
Prunus ilicifolia
*Prosopis chilensis
Lotus strigosus
Lotus scoparius
Astragalus didymocarpus
Astragalus oxyphysus
*Rhamnus californica
*Ceanothus cuneatus
Ceanothus Ferrisae
Mentzelia gracilienta
Oenothera micrantha var. Jonesii
Oenothera contorta var. strigulosa
Lomatium dasycarpum
Arbutus Menziesii[4]
*Arctostaphylus glauca
Forestiera neomexicana

*Fraxinus dipetala
Hugelia pluriflora
Hugelia filifolia var. typica
Gilia multicaulis
Pholistoma aurita
Pholistoma membranacea
Lemmonia californica
Eucrypta chrysanthemifolia
Emmenanthe penduliflora and var. rosea
Phacelia ramosissima var. suffrutescens
Pectocarya linearis var. ferocula
Pectocarya setosa
Amsinckia Eastwoodae
Amsinckia vernicosa
Amsinckia tesselata
Cryptanthe Clevelandii
Cryptanthe corollata
Salvia mellifera
Salvia carduacea
Mimulus androsaceus
Lonicera Johnstoni
Lonicera interrupta
Nemacladus ramosissimum
Stephanomeria exigua var. coronaria
Malacothrix Coulteri
Malacothrix Clevelandii
Gutierrezia californica
Stenotopsis linearifolius
Eastwoodia elegans
Chrysothamnus nauseosus var. mohavensis
Chrysopsis villosa vars.
Lessingia germanorum var. parvula
Corethrogyne filaginifolia
Stylocline filaginifolia
Coreopsis calliopsidea
Coreopsis Douglasii
Coreopsis hamiltonii
Hemizonia Kelloggii
Hemizonia fasciculata
Monolopia major
Eriophyllum Wallacei
Artemisia californica
Senecio Breweri
*Lepidospartum squamatum

4 Species listed in "northern" element as well; presumably with a southern origin, migrating north in the Eocene to become elements in the Miocene redwood forest and to migrate south again.

GREAT BASIN AND DESERT DERIVATIVES

Many Great Basin and desert derivatives of the Sonoran floristic element are entirely restricted, in California, to the desert areas. Some, however, have a sufficiently broad range of climatic tolerance to allow for limited establishment westward beyond the desert areas, or they may be replaced there by closely related species. The major area where such species occur is the foothill zone which encircles the head of the San Joaquin Valley (Fig. 3). This includes the Kern basin area at the southwestern margin of the Sierra Nevada, the northern base of the Tehachapi Mountains, and the eastern base of the inner South Coast Ranges (Temblor and Diablo ranges). The species of this group which reach the Diablo Range are usually restricted to a few isolated localities. In the Mount Hamilton Range these species form a small but highly distinctive unit of the flora. A list of them is given below. Their Great Basin and desert affinities are evident by even a casual perusal of the list.

Juniperus californicus	**Phacelia Fremontii*
Puccinellia simplex	*Lemmonia californica*
Allium fimbriatum	*Emmenanthe penduliflora* var. *rosea*
*Allium peninsulare var. *crispum*	*Pectocarya setosa*
Allium lacunosum	*Amsinckia grandiflora*
Chorizanthe uniaristata	**Amsinckia vernicosa*
Chorizanthe Clevelandii	**Amsinckia tessellata*
*Chorizanthe perfoliata	*Cryptanthe nevadensis* var. *rigida*
Eriogonum angulosum	**Plagiobothrys arizonicus*
*Calyptridium monandrum	*Salvia carduacea*
Streptanthus Coulteri var. *Lemmonii*	**Monardella Breweri*
Streptanthus lilacinus	*Malacothrix Coulteri*
Tropidiocarpum capparideum	**Eastwoodia elegans*
*Prosopis chilensis	**Chrysothamnus nauseosus* var. *mohavensis*
*Astragalus oxyphysis	**Coreopsis calliopsidea*
Euphorbia ocellata	**Coreopsis hamiltonii*
*Malvastrum Parryi	**Coreopsis Douglasii*
*Oenothera deltoides var. *cognata*	**Hemizonia pungens*
*Oenothera decorticans var. *typica*	*Madia radiata*
Forestiera neomexicana	**Eriophyllum Wallacei*
*Hugelia pluriflora	*Senecio Breweri*
*Pholistoma membranacea	**Lepidospartum squamatum*

* Northern limit in Mount Hamilton Range.

These species are restricted almost entirely to the eastern margin of the Mount Hamilton Range, although a few are found in the interior (*Juniperus californicus, Chorizanthe perfoliata, Forestiera neomexicana, Coreopsis Douglasii, Eriophyllum Wallacei, Senecio Breweri, Lepidospartum squamatum*) or on the summit areas of the western crest (*Calyptridium monandrum, Chrysothamnus nauseosus* var. *mohavensis, Coreopsis hamiltonii*), but none reaches the more mesophytic western slopes of the range.

Over half of the species listed have their known northern limit in the Mount Hamilton Range. This large percentage is understandable on the basis of the presumed origin, migrational history, and environmental requirements of the group as a whole. As to origin, the Miocene Tehachapi flora described by Axelrod (1939) possesses fossil representatives of modern species whose distributional pattern coincides with that of the species in question—the Great Basin, desert regions of California, southern Sierra Nevada, and inner South Coast Ranges. On the basis of their fossil equivalents, the modern species are considered by Axelrod to have arisen in the Sierra Madre of northern Mexico, and thus they are members of the Sonoran (Sierra Madrean) floristic element. It is logical to assume that, even when fossil equivalents are lacking, modern species with a similar distribution to the above have a similar origin.

Considering migrational history next, the Walker Pass (oral communication, H. L. Mason in 1939), and to a lesser degree the Tehachapi Pass and the Tejon Pass (Parish, 1903, 1930), form the principal gateways for the migration of the typically desert species from the western margin of the Mohave Desert into the foothill zone at the head of the San Joaquin Valley (Fig. 3). The elevation here is much lower (400-1000 feet) than in the western part of the Mohave Desert (2700-3000 feet), although the temperature extremes are slightly less and the precipitation is slightly more (5.72 inches at Bakersfield, 4.86 inches at Mohave). The environment is sufficiently desert-like, however, to permit the establishment of certain desert species. According to Bauer (1930), the northern base of the Tehachapi Mountains at the extreme head of the San Joaquin Valley constitutes a "semi-desert." Paleobotanical evidence indicates that migration into these areas did not become definitely established until the late Pliocene, and presumably most of it has occurred subsequent to the post-Pleistocene development of a sufficiently warm, arid climate in these extra-desert areas.

Climatic conditions comparable to those at the head of the San Joaquin Valley prevail along the eastern margin of the inner South Coast Ranges, permitting the northward migration and establishment of some of the more tolerant desert species. The climate undergoes a gradual but progressive northward moderation, however, the temperature extremes lessening and the rainfall increasing, and coincident with these climatic changes there is a gradual dropping out of the desert species, so that the farther north one progresses, the fewer are found. The eastern side of the Mount Hamilton Range represents the last extensive area favorable to typically desert species. This accounts for the large number of these species which have their northern outpost in the Mount Hamilton Range. A few reach their northern limit on Mount Diablo or in the region east of Mount Diablo (*Eriogonum angulosum, Forestiera neomexicana, Salvia carduacea, Malacothrix Coulteri, Madia radiata, Senecio Breweri*), and at least four (*Juniperus californicus, Allium fimbriatum, Chorizanthe Clevelandii*, and *Lemmonia californica*) are found in the "hot-summer" climatic zone which occurs in the vicinity of Lake County of the North Coast Ranges. Thus the extra-desert distributional pattern of these desert species can be likened to a gradually tapering inverted V, the base of which is curved and lies in the northern foothills of the Tehachapi Mountains, while the extreme apex extends approximately to Lake County in the North Coast Ranges. Climate is the major factor controlling the extra-desert migration and establishment of these species, as it is likewise the major controlling factor in their gradual northward restriction.

In certain of the species, however, the edaphic factor appears to exert a secondary influence on distribution, as in *Allium fimbriatum* and *Emmenanthe penduliflora* var. *rosea*. In the Mount Hamilton Range both of these species are restricted to the serpentine areas of the Red Mountains. The limited occurrence of *Allium fimbriatum* in the inner South Coast Ranges and its long "jump" from the Mount Hamilton Range to the serpentine areas of Lake

County in the North Coast Ranges may be partly the result of an edaphic limitation to serpentine soils. *Allium lacunosum* is similarly, although less strongly, influenced in its distribution by the occurrence of serpentine soils.

Some of the Great Basin and desert derivatives listed do not occur in the desert areas, but are restricted to a larger or smaller portion of the foothill zone of the San Joaquin Valley. These are *Chorizanthe uniaristata, C. Cleve-landii, Streptanthus Coulteri* var. *Lemmonii, S. lilacinus, Tropidiocarpum capparideum, Astragalus oxyphysus, Malvastrum Parryi, Oenothera decorticans* var. *typica, O. deltoides* var. *cognata, Hugelia pluriflora, Emmenanthe penduliflora* var. *rosea, Amsinckia grandiflora, Eastwoodia elegans, Senecio Breweri, Coreopsis hamiltonii,* and *Coreopsis Douglasii.* They are associated with typically desert species both distributionally and taxonomically, however, as the following examples suggest:

Extra-desert species	Typically desert species
Streptanthus Coulteri var. *Lemmonii*	*Streptanthus Coulteri*
Oenothera decorticans var. *typica*	*O. decorticans* (vars. *rutila, desertorum, condensata*)
Oenothera deltoides var. *cognata*	*O. deltoides*
Emmenanthe penduliflora var. *rosea*	**Emmenanthe penduliflora*
Coreopsis hamiltonii	*C. Bigelovii*
Coreopsis Douglasii	**C californica*

* = Also well developed in cismontane southern California.

Coreopsis hamiltonii–C. Bigelovii and *C. Douglasii–C. californica* form two such instances of close distributional and taxonomic relationships (Shar-smith, 1938). *Coreopsis hamiltonii* has somewhat less arid and hot climatic requirements than the typically desert species, *C. Bigelovii,* and is known only from the Mount Hamilton Range, although its morphology suggests it to be closely related to and a derivative of the desert species, *C. Bigelovii.* Similarly, *C. Douglasii* is known only from the inner South Coast Ranges, but it is closely related to and considered to have been derived from *C. californica* of the desert and cismontane areas of southern California.

CISMONTANE SOUTHERN CALIFORNIA DERIVATIVES

Another group of derivatives of the Sonoran floristic element is closely related to the typically desert species just considered. It consists of species which have their centers of distribution in northern Baja California and cis-montane southern California or which have their affinities in these areas, but which are presumed to have been derived from the Sonoran floristic element. They occur northward in the South Coast Ranges approximately to central California and may occur northward in the southern Sierra Nevada as well. A representative list of these Mount Hamilton Range species is:

*Selaginella Bigelovii
Pinus Coulteri
Stipa pulchra
Stipa lepida
*Allium Parryi
*Muilla serotina
*Calochortus invenustus
Calochortus clavatus
Chorizanthe polygonoides
Eriogonum virgatum
*Eriogonum saxatile
*Eriogonum fasciculatum var. foliolosum
*Calyptridium Parryi
*Delphinium Parryi
Papaver heterophyllum
Tropidiocarpum gracile
*Lithophragma Cymbalaria
Ribes aureum var. gracillimum
Ribes malvaceum
*Ribes quercetorum
Ribes amarum
*Ribes speciosum
Prunus ilicifolia
Lotus strigosus
Lotus scoparius
Astragalus didymocarpus
Mentzelia gracilienta
Oenothera micrantha var. Jonesii

Oenothera contorta var. strigulosa
Lomatium dasycarpum
Arctostaphylos glauca
Hugelia filifolia (var. typica)
Gilia multicaulis
Pholistoma aurita
Eucrypta chrysanthemifolia
*Phacelia ramosissima var. suffrutescens
Pectocarya linearis var. ferocula
Amsinckia Eastwoodae
Cryptanthe Clevelandii
*Cryptanthe corollata
Salvia mellifera
*Mimulus androsaceus
*Lonicera Johnstoni
*Nemacladus ramosissimum
*Stephanomeria exigua var. coronaria
Malacothrix Clevelandii
Gutierrezia californica
Stenotopsis linearifolius
Chrysopsis villosa vars.
Lessingia germanorum var. parvula
Corethrogyne filaginifolia
Stylocline gnaphaloides
Hemizonia Kelloggii
Hemizonia fasciculata
Artemisia californica

* = Northern limit in or approximately at the Mount Hamilton Range.

As with the typically desert species, climate is assumed to be the major limiting factor in the northward distribution of these typically cismontane southern California species. They form part of a larger number of species whose climatic tolerance entirely limits them to cismontane southern California. The species being considered are assumed to have a sufficiently broader range of climatic tolerance, however, to allow for their limited northward establishment. In the Coast Ranges this is mainly in the inner South Coast Ranges, although some may occur on the eastern slopes of the outer South Coast Ranges. In the Mount Hamilton Range, they may be found in all regions except the more highly mesophytic portions of the western slopes, although the greater number occurs only in the interior of the range.

As with the typically desert species, the Mount Hamilton Range forms the northern limit for many of these species, probably because it is the last site climatically favorable to their northward establishment. Several species are found northward beyond the Mount Hamilton Range only to Mount Diablo (Pinus Coulteri, Ribes amarum, Arctostaphylos glauca, Eucrypta chrysanthemifolia, Salvia mellifera, Stenotopsis linearifolium), while a few are found in the zone of "hot-summer" climate which occurs in the vicinity of Lake County in the inner North Coast Ranges (Papaver heterophyllum, Hugelia filifolia var. typica, Gilia multicaulis).

Again, as with the typically desert species, the edaphic factor appears to represent a limiting influence secondary to climate in the northward distribution of some of these typically cismontane southern California species. For example, in the Mount Hamilton Range, *Allium Parryi*, like *A. fimbriatum* of the preceding group, occurs only in the serpentine areas of the Red Mountains, and its highly restricted occurrence in the South Coast Ranges may be partly the result of an edaphic limitation to serpentine soils.

CENTRAL COAST RANGE ENDEMIC ELEMENT

The Mount Hamilton Range is characterized by a fairly high percentage (approximately 13 per cent) of species and varieties endemic to the Coast Ranges of California, of which about 3 per cent are limited to the central Coast Ranges. The latter are highly restricted in distribution, and are spoken of here as "narrow" endemics, although such a classification is, of course, relative. On the basis of their morphology, distribution, and taxonomic affinities, many of the Coast Range endemics of the Mount Hamilton Range can be considered as depleted species, once of wider distribution, which have an origin and history in common with either the northern or the Sonoran floristic elements. Where the evidence indicated this, such Coast Range endemics were included in one or the other of these elements. Thus, under the northern element, these Coast Range endemics were included: *Claytonia gypsophiloides, Berberis pinnata, Ribes sanguineum* var. *glutinosum*, and *Amelanchier alnifolia* var. *subintegra*. Under the Sonoran element, these Coast Range endemics were included: *Chorizanthe Clevelandii, C. uniaristata, Eriogonum nudum* var. *auriculatum, Streptanthus Coulteri* var. *Lemmonii, Oenothera contorta* var. *strigulosa, Lonicera interrupta, Coreopsis hamiltonii*, and *C. Douglasii*.

Many of the highly restricted and discontinuously distributed central Coast Range endemic species have an insular origin; that is, they are species which, by one means or another, became established on the Coast Range archipelago, where their continued isolation led to genetic uniformity and to distinction from the ancestral stocks. The origin and nature of insular species of this type is discussed by Stebbins (1942) in a paper which considers the problem of endemic species from the genetic viewpoint. The Mount Hamilton Range species of presumed insular origin are listed below, together with a condensed statement of the areas they occupy:

Allium falcifolium.[5] Eastern summits of Santa Cruz Mountains; Mount Hamilton Range; Mount Diablo; Sonoma County to southern Oregon.
Fritillaria folcata. Mount Hamilton Range; San Carlos Range.
Salix Breweri. San Carlos Range; Mount Hamilton Range; Napa, Lake, and Colusa counties of North Coast Ranges.
Eriogonum Covilleanum. San Carlos Range; Mount Hamilton Range.
Arenaria Douglasii var.[6] Mount Hamilton Range.
Delphinium californicum var. *interius.* Mount Hamilton Range; Mount Diablo.
Arabis Breweri. Santa Cruz Mountains; Mount Hamilton Range; Mount Diablo; Lake and Napa to Siskiyou counties in North Coast Ranges.

---

5 Not strictly a central Coast Range endemic species, and not in its entirety an insular species. Its segregate, *Allium Breweri* (Mount Diablo, Mount Hamilton Range, eastern summits of Santa Cruz Mountains), here considered conspecific with *A. falcifolium*, appears to have an insular history and may be genetically distinct.

6 See page 337.

Streptanthus albidus. Mount Hamilton Range; Mount Diablo.

Streptanthus Breweri. San Carlos Range; Mount Hamilton Range; Glenn and Lake counties in North Coast Ranges.

Streptanthus callistus. Mount Hamilton Range.

Linum Clevelandii. Red Mountains of Mount Hamilton Range; Napa, Lake, and Mendocino counties in North Coast Ranges.

Sedella pentandra. Gavilan Range; San Carlos Range; Mount Hamilton Range; Lake County of North Coast Ranges.

Trifolium dichotomum var. *turbinatum.* Mount Hamilton; Mount St. Helena and Mount Tamalpais in North Coast Ranges.

Lotus rubriflorus. Mount Hamilton Range.

Lathyrus Bolanderi var. *quercetorum.* Mount Hamilton Range; Mount Diablo.

Clarkia Breweri. Mount Hamilton Range; eastern side Santa Cruz Mountains; San Carlos Range; Mayacamas Range, Sonoma County, in North Coast Ranges.

Sanicula saxatilis. Mount Hamilton; Mount Diablo.

Garrya Congdoni.[7] San Carlos Range; Mount Hamilton Range; Tehama and Lake counties in North Coast Ranges.

Collomia diversifolia. Mount Hamilton Range; Lake County in North Coast Ranges.

Navarretia Abramsii. Mount Hamilton Range; eastern side Santa Cruz Mountains; Lake County in North Coast Ranges.

Linanthus ambiguus. Mount Hamilton Range; eastern slopes Santa Cruz Mountains; Mount Diablo.

Phacelia Breweri. Gavilan Range; San Carlos Range; eastern side Santa Cruz Mountains; Mount Hamilton Range; Mount Diablo.

Phacelia phacelioides. Mount Hamilton Range; Mount Diablo.

Acanthomintha lanceolata. Eastern slopes Santa Cruz Mountains; Mount Hamilton Range.

Castilleja Roseana. San Carlos Range; Mount Hamilton Range.

Campanula exigua. Mount Hamilton Range; Mount Diablo.

Eriophyllum Jepsonii. San Carlos Range; Mount Hamilton Range; Mount Diablo.

Cirsium campylon. Mount Hamilton Range.

The origin and migrational history of these insular species is correlated with the geologic history of the Coast Ranges, and with the edaphic and climatic factors of the habitats which have been imposed upon them in consequence of their geologic history. As to geologic history first, not only are these endemics restricted to the central Coast Ranges, but their strikingly interrupted pattern of distribution coincides almost exactly with those porions of the Coast Ranges which are composed of the Franciscan rock series (Fig. 2).[8] The present exposure of Franciscan rocks in the South Coast Ranges, and presumably in the North Coast Ranges as well, corresponds to those areas which have been insular land masses since post-Jurassic times. These species, then, may be viewed as insular species restricted essentially to the emergent areas of Franciscan rock they occupied during the various periods of coastal inundation, particularly during the Pliocene-Pleistocene revolution when the Coast Ranges represented an archipelago system.

---

7 Also in central Sierra Nevada in areas of the Mariposa rock series, a series homologous to the Franciscan series in the Coast Ranges.

8 The observations recorded by Greene (1893b) of a north-south trend in the distribution of rare central California species are of interest here.

Jepson (1925) calls species of this sort insular relicts. He lists *Allium Breweri*,[9] *Streptanthus hispidus*,[10] *Sanicula saxatilis*, *Phacelia phacelioides*, possibly *Phacelia Breweri*, and *Campanula exigua* as "survivals on the mountains peaks of the central Coast Ranges, that is, the summit areas of Mount St. Helena, Mount Diablo, Mount Hamilton and Loma Prieta." As already discussed, however, not only the mountain peaks, but most of the surface area of the Mount Hamilton Range, the area now covered by Franciscan rocks, escaped invasion by the sea; correlated with this we find these species not localized on the mountain summits, but restricted to areas of Franciscan rock. Thus *Phacelia Breweri* and *Campanula exigua* occur across the Mount Hamilton Range in localized colonies from altitudes of 1000 to 4200 feet. *Phacelia phacelioides* occurs in the interior at 2000 feet altitude. Also, if these species are true insular species, their distribution has always been restricted to the island areas occupied, and they are not relicts in the sense that they were once more widespread.

These insular species are insular not only by virtue of their restriction to areas which once represented actual islands, but they are also insular in the sense that they are edaphically isolated. Their edaphic isolation rests upon the fact that they are restricted to areas of Franciscan rocks. As we have seen, the island areas and the Franciscan rock areas are coincident, and it was the actual island areas exposed during marine inundations which primarily imposed upon these species their limitation to Franciscan rocks. Secondarily, however, the distinctive soils of the Franciscan rocks must also have been a factor in the differentiation of the original insular plant populations from the continental ancestors; and later, when the seas withdrew, these soils must also have been a factor in preventing any appreciable migration beyond the original island boundaries.

Particularly in the case of serpentinophilous species, the evidence for edaphic control is striking. *Fritillaria folcata* has been found only on serpentine talus, in the Red Mountains of the Mount Hamilton Range and on San Benito Peak of the San Carlos Range. Related species, *F. glauca* and *F. Purdyi*, are limited species found only on serpentine in the Lake County area and somewhat north in the North Coast Ranges. *Cirsium campylon* is found only in the Mount Hamilton Range and only on serpentine soils. Its only close relative, *C. fontinale*, is an insular endemic of the outer Coast Ranges, restricted entirely to serpentine. *Salix Breweri*, *Streptanthus Breweri*, *Linum Clevelandii*, *Garryi Congdoni*, and *Collomia diversifolia* are found only on serpentine in the Mount Hamilton Range, and as far as is known, they are obligate to serpentine elsewhere. These latter species are not found between the serpentine of Lake County and the Mount Hamilton Range, and the edaphic factor evidently exerts an important control in their present discontinuous and limited distribution.

---

9 See *Allium falcifolium*, preceding list.

10 Only on Mount Diablo according to J. L. Morrison (oral communication in 1939).

Other of the central Coast Range insular endemics show a similar restricted and disjunct distribution. Although not limited to serpentine, *Clarkia Breweri*, *Navarretia Abramsii*, and *Sedella pentandra*, for example, have the same distribution as the obligate serpentine species, and do not occur between the Mount Hamilton Range and the Lake County Franciscan area of the inner North Coast Ranges (vicinity of Clear Lake, Fig. 2). This latter area includes southern Lake County and adjacent portions of Napa, Sonoma, Glenn, and Colusa counties. The distinction of the flora of this area from surrounding portions of the North Coast Ranges was pointed out by Jepson many years ago (1893).

North of this Lake County Franciscan area in the North Coast Ranges, the Franciscan rocks continue for a considerable distance, and the question arises as to why most of the disjunct species which occur both in the Mount Hamilton Range and in the Lake County area do not continue northward in the inner North Coast Ranges. Both climatic and edaphic controls are apparently involved. Edaphic control seems to be due to a gradual change north of Lake County in the nature of the Franciscan rocks, and the resultant change in character of the soils produced, whereas in the Lake County area where the species under consideration appear, the Franciscan rocks are much like those of the Mount Hamilton Range and produce the same types of soils. As to climate, most of the disjunct species seem to be limited to areas of hot-summer "olive" climate, and this is the type of Mediterranean climate found both in the Lake County area and throughout the interior of the Mount Hamilton Range, whereas north of Lake County the inner North Coast Ranges are predominantly in a cool-summer "heather" climate.

This restriction to areas of hot-summer "olive" climate also appears to be the reason why many of the disjunct species, including the obligate serpentine dwellers, occur both in the Lake County area and in the Mount Hamilton Range, but "skip" the intervening areas of Franciscan rock which appear on the summit of Mount Diablo and in the Contra Costa Hills. The summit of Mount Diablo has a cool-summer "heather" climate like that of the inner North Coast Ranges north of Lake County, while the Franciscan area of the Contra Costa Hills approaches a maritime climate due to its position near San Francisco Bay.

Some of the endemic species (*Arabis Breweri*, *Linanthus ambiguus*, *Phacelia phacelioides*, *P. Breweri*, and *Campanula exigua*) are limited to a cool-summer "heather" climate or have a climatic tolerance sufficient to account for their presence in both cool-summer and hot-summer types of Mediterranean climate. These species occur both on Mount Diablo and in the Mount Hamilton Range. The very distinctive and extremely narrow endemic, *Sanicula saxatilis*, is limited to several isolated colonies on the upper slopes of Mount Diablo and Mount Hamilton in the cool-summer zone of the higher altitudes. Climate also furnishes a clue why most of the Mount Hamilton Range insular species occur only in the inner Coast Ranges and not in the outer, for the climate of the outer Coast Ranges is considerably more maritime. Closely related but distinct species may occur on the insular areas of the outer and inner Coast Ranges. *Campanula exigua*, for example, is represented by a closely related species, *C. angustiflora*, on the Franciscan rocks of Mount Tamalpais (outer North Coast Ranges) and Mount St. Helena (inner North Coast Ranges). The morphological divergence of these two units may be explained in part by the climatic divergence in the different insular areas occupied. Similar cases are those of *Cirsium campylon* and *Castilleja Roseana*. *Cirsium campylon*, an obligate serpentine species, has been discussed. *Castilleja Roseana*, found only in the Mount Hamilton Range and the San Carlos Range, finds its closest relative in the common and strictly maritime species, *C. latifolia*, of the outer Coast Ranges.

In review, the geologic history and the edaphic and climatic conditions of the Franciscan land masses have been considered as interacting factors involved in varying degree in the origin and distributional history of the central Coast Range species restricted to these areas. From the viewpoint of geologic history of the areas occupied, these species have been designated as insular endemics. It is assumed that these species or their precursors migrated into the central Coast Ranges from both northern and southern sources during the early Tertiary. Their subsequent long and isolated history on the insular land masses of the Coast Ranges, under the influence of distinctive edaphic conditions, has resulted in differentiation from the parental stocks, so that they now represent a separate unit in the California flora. Their long isolation resulted, especially for those species restricted to a single island area, in relatively non-plastic species with specialized edaphic requirements and a rather narrow range of climatic tolerance, so that in the post-Pleistocene period of coastal emergence they were unable to migrate appreciably beyond the Franciscan land blocks.

Acknowledgments

Acknowledgment is expressed to all who have given the writer assistance, in particular: staff members of the various herbaria consulted, especially those of the California Academy of Sciences, Stanford University, University of California, and Vegetative Type Map Herbarium of the California Forest and Range Experiment Station, both for loaning specimens and for responding to requests for information; the various botanists who have verified, corrected, or given the determinations in certain groups of plants; R. F. Hoover, for the loan of his collections from the eastern margin of the Mount Hamilton Range; and Annetta M. Carter, for invaluable assistance in certain phases of the field work. Finally, sincere appreciation is given to H. L. Mason, under whose guidance this work was done, for his continued, stimulating interest and assistance in its progress.

REFERENCES

Anderson, F. M. 1905—A stratigraphic study in the Mount Diablo Range of California. Proc. Calif. Acad. Sci. ser. 3, Geol. **2**:156-248.

————1908—A further stratigraphic study in the Mount Diablo Range of California. Proc. Calif. Acad. Sci. ser. 3, Geol. **3**:1-40.

Anderson, R. and R. W. Pack. 1915—Geology and oil resources of the west border of the San Joaquin Valley north of Coalinga; California. U. S. Geol. Surv. Bull. **603**:1-220.

Axelrod, D. I. 1938—The stratigraphic significance of a southern element in later Tertiary floras of western America. Journ. Wash. Acad. Sci. **28**:313-322.

————1939—A Miocene flora from the western border of the Mohave Desert. Carn. Inst. Wash. Publ. **516**:1-129.

BABCOCK, E. B. AND G. L. STEBBINS, JR. 1938—The American species of *Crepis*. Carn. Inst. Wash. Publ. **504**:1-189.

BAUER, H. L. 1930—Vegetation of the Tehachapi Mountains, California. Ecology **11**:263-280.

BRAUN-BLANQUET, J. 1932—Plant Sociology. McGraw Hill Book Co., New York.

BREWER, W. H. 1930—Up and down California in 1860-1864. Edited by F. P. Farquhar. Yale Univ. Press. New Haven, Conn.

BYERS, H. R. 1930—Summer sea fogs of the central California Coast. Univ. Calif. Publ. Geog. **3**:291-338.

CALIFORNIA STATE MINING BUREAU. 1893—Eleventh annual report of state mineralogist, 374-375.

———1915—Twenty-third annual report of state mineralogist, 201-208.

CHANEY, R. W. 1936—The succession and distribution of Cenozoic floras around the northern Pacific Basin, pps. 55-85 in Essays in Geobotany. Edited by T. H. Goodspeed. Univ. Calif. Press. Berkeley.

CLARK, B. L. 1925—Tectonics of the Vallé Grande and surrounding areas. Bull. Geol. Soc. Am. **36**:165 (abstract).

———1927—Studies in the tectonics of the Coast Ranges. Bull. Geol. Soc. Am. **38**:163.

———1929—Tectonics of the Vallé Grande of California. Bull. Amer. Assoc. Petrol. Geol. **13**:199-238.

———1930—Tectonics of Coast Ranges of middle California. Bull. Geol. Soc. Am. **41**:747-828.

———1935—Tectonics of the Mount Diablo and Coalinga areas, middle Coast Ranges of California. Bull. Geol. Soc. Am. **46**:1025-1078.

COOPER, W. S. 1922—The broad sclerophyll vegetation of California. Carn. Inst. Wash. Publ. **319**:1-124.

DAVIS, E. F. 1918a—The Franciscan sandstone. Univ. Calif. Publ. Geol. **11**:1-44.

———1918b—The radiolarian cherts of the Franciscan group. Univ. Calif. Publ. Geol. **11**:235-432.

FAIRBANKS, H. W. 1894—Review of our knowledge of the geology of the California Coast Ranges. Bull. Geol. Soc. Am. **6**:71-102.

———1898—Geology of a portion of the southern Coast Ranges. Journ. Geol. **6**:551-576.

GORDON, A. AND C. B. LIPMAN. 1926—Why are serpentine and other magnesian soils infertile? Soil Sci. **22**:291-302.

GREENE, E. L. 1893a—The vegetation of the summit of Mount Hamilton. Erythea **1**:77-97.

———1893b—The vegetation of Mount Diablo. Erythea **1**:166-179.

HELLER, A. A. 1907—Botanical exploration in California, season of 1907. 1. The Coast Region. Mühlenbergia **2**:269-338.

JENKINS, O. P. 1938—Geologic Map of California. State of California, Dept. of Natural Resources, Div. of Mines.

JEPSON, W. L. 1893—The mountain region of Clear Lake. Erythea **1**:10-16.

———1925—A manual of the flowering plants of California. Assoc. Students Store, Univ. Calif., Berkeley.

LAWSON, A. C. 1893—The post-Pliocene diastrophism of the coast of southern California. Univ. Calif. Publ. Geol. **1**:115-160.

LAWSON, A. C. AND C. PALACHE. 1902—The Berkeley Hills, a detail of Coast Range geology. Univ. Calif. Publ. Geol. **2**:349-450.

McADIE, A. G. 1903—Climatology of California. U. S. Dept. Agri. Weather Bureau Bull. **50**:1-270.

———1914—The rainfall of California. Univ. Calif. Publ. Geog. **1**:127-240.

PARISH, S. B. 1903—A sketch of the flora of southern California. Bot. Gaz. **36**:203-222, 259-279.

———1930—Vegetation of the Mohave and Colorado deserts of southern California. Ecology **11**:481-499.

PIRSSON, L. V. 1926—Rocks and rock minerals. New York.

REED, R. D. 1933—Geology of California. Am. Assoc. Petrol. Geol. Tulsa, Okla.

RUSSELL, R. J. 1926—The climates of California. Univ. Calif. Publ. Geog. **2**:73-84.

SHARSMITH, H. K. 1936—The genus *Sedella*. Madroño 3:240-248.
————1938—The native Californian species of the genus *Coreopsis* L. Madroño 4:209-231.
————1939—A new species of *Cirsium* from California. Madroño 5:85-90.
————1940—Further notes on the genus *Sedella*. Madroño 5:192-196.
————1941—A new species of *Lotus* from the Mount Hamilton Range, California. Madroño 6:56-58.
STEBBINS, G. L., JR. 1938—The western American species of *Paeonia*. Madroño 4:252-260.
————1942—The genetic approach to problems of rare and endemic species. Madroño 6:241-258.
TAFT, J. A. 1935—Geology of Mount Diablo and vicinity. Bull. Geol. Soc. Am. 46:1079-1100.
TALIAFERRO, N. L. 1943—Geologic history and structure of the central Coast Ranges of California. State of California, Dept. of Natural Resources, Div. of Mines, Bull. 118(pt. 2):119-162.
TEMPLETON, E. C. 1913—General geology of the San Jose and Mount Hamilton quadrangles. Bull. Geol. Soc. Am. 24:96 (abstract).
TURNER, H. W. 1891—The geology of Mount Diablo, California. Bull. Geol. Soc. Am. 2:383-414.
VICKERY, F. P. 1925—The structural dynamics of the Livermore region. Journ. Geol. 33:608-628.
WILLIS, R. 1925—Physiography of California Coast Ranges. Bull. Geol. Soc. Am. 36:640-678.
WHITNEY, J. D. 1865—Geological survey of California, Vol. 1. Caxton Press, Philadelphia.

Annotated List of Vascular Species

The species of vascular plants known to occur in the Mount Hamilton Range are listed in the following pages. Careful taxonomic evaluation of these species was considered vital, for upon such evaluation rests the validity of the conclusions advanced to explain the geographical affinities of the flora. Determinations were checked by specialists wherever possible. For a number of the species, especially those of limited distribution, a considerable sum of new information, both morphological and distributional, was accumulated during the writer's field and herbarium work. Some of it has already been published by the writer and others, but any previously unpublished new material is included in this list. There is only one new entity, a variety, described herein (p. 337), but four species have been described in separate papers: *Lotus rubriflorus* (Sharsmith, Madroño 6:56. 1941), *Sedella pentandra* (Sharsmith, Madroño 3:240. 1936), *Cirsium campylon* (Sharsmith, Madroño 5:85. 1939), and *Streptanthus callistus* (Morrison, Madroño 4:205. 1938).

Specimens from the following herbaria are cited in the annotated list: California Academy of Sciences (CA), Gray Herbarium (G), Missouri Botanical Garden (Mo), Royal Botanic Gardens, Kew (K), Pomona College (P), Stanford University (S), State College of Washington (WSC), University of California (UC), and Vegetative Type Map Herbarium of the California Forest and Range Experiment Station (VTM).

In using monographic treatments it was the policy, in general, to accept the taxonomic treatment of the monographer concerned. This accounts for the interchangeable use, in some cases, of the categories of "variety" and "subspecies." Synonyms (in italics and parentheses) are usually given only if they represent binomials accepted in current manuals or in other botanical literature which refers to the Mount Hamilton Range. The habitat and distribution of each species is given, and distribution is so stated as to give the occurrence of the species *across* the range from west to east, for, as discussed, the major floristic changes are met with latitudinally. Specimens are cited only for range extensions, rare species, or species involving a taxonomic problem. Types or isotypes which have been examined by the writer are designated by the symbol !. Introduced species are indicated by an asterisk. New names or combinations are printed in bold face. All localities cited are included in Figure 1.

POLYPODIACEAE

CYSTOPTERIS FRAGILIS (L.) Bernh. (*Filix fragilis* (L.) Gilib.) Sheltered ravines and wooded slopes, west side Mount Hamilton.

POLYPODIUM VULGARE L. var. KAULFUSSII (D. C. Eat.) Fern. (*P. californicum* Kaulf.) Rock crevices and wooded slopes, west side Mount Hamilton.

POLYSTICHUM MUNITUM (Kaulf.) Presl. (*Aspidium munitum* Kaulf.) Wooded slopes, west side of range.

DRYOPTERIS ARGUTA (Kaulf.) Watt. (*Aspidium rigidum* var. *argutum* D.C. Eat.) Rocky wooded areas, west side and interior of range.

WOODWARDIA FIMBRIATA Sm. (*W. Chamissoi* Brack.; *W. radicans* of authors.) North base Mount Hamilton.

PITYROGRAMMA TRIANGULARIS (Kaulf.) Maxon. (*Gymnogramma triangularis* Kaulf.) Occasional across range on wooded slopes, rock ledges and crevices, and talus.

ADIANTUM JORDANI Müll. (*A. emarginatum* sensu D. C. Eat., non Bory.) Sheltered rock ledges or moist wooded slopes across range, uncommon.

CHEILANTHES GRACILLIMA D. C. Eat. Exposed rock ledges near summit Mount Hamilton.

CHEILANTHES INTERTEXTA (Maxon) Maxon. (*C. Covillei* Maxon subsp. *intertexta* Maxon.) Occasional on exposed rock ledges, west side and interior of range.

PELLAEA MUCRONATA (D. C. Eat.) D. C. Eat. (*P. ornithopus* Hook.) Occasional across range on rocky slopes and ledges.

PELLAEA ANDROMEDIFOLIA (Kaulf.) Fee. Occasional across range on rocky slopes.

MARSILEACEAE

MARSILEA VESTITA Hook. and Grev. Still water, ponds or streams.

EQUISETACEAE

EQUISETUM LAEVIGATUM A. Br. Margins of Santa Isabella Creek.

EQUISETUM ARVENSIS L. North base Mount Hamilton; bog.

SELAGINELLACEAE

SELAGINELLA BIGELOVII Underw. Occasional across range on exposed rock crevices and slopes.

ISOETACEAE

ISOETES HOWELLII Engelm. Mount Day Ridge; pond.

PINACEAE

PINUS PONDEROSA Dougl. Grove of ten or twelve trees growing among *P. Coulteri*, east slope Mount Hamilton near summit, *H. K. Sharsmith 524* (UC). Only two small and isolated groves of *Pinus ponderosa* are known on the mountain, the one cited and another a short distance below on the same slope. Greene (Erythea 1: 97. 1893) states, "Fine groves of it [*P. ponderosa*] are seen near the summits of high ridges not far away." Apparently Greene mistook *P. Coulteri* for this species, for he does not list *P. Coulteri*, although it occurs on the summit ridges of Copernicus Peak, a portion of the

Mount Hamilton summit. *Pinus Coulteri* is the abundant pine at altitudes which might support either species.

PINUS COULTERI Don. Common on summit ridges of west crest of range: Santa Isabella Peak, Mount Hamilton, Packard Ridge, Valpé Ridge, Mount Day Ridge, Mount Lewis, at altitudes of 3000 to 4000 feet, growing most abundantly or exclusively on the eastern slopes of these ridges. This species, mainly of southern California, reaches its northernmost extension in the Mount Hamilton Range except for two colonies in the vicinity of Mount Diablo.

PINUS SABINIANA Dougl. Across range on exposed slopes, ridge tops, gravelly flats, or flood beds of intermittent streams. *Pinus Sabiniana* mingles with *P. Coulteri* at the lower levels of the latter species, or the two sometimes occupy the higher ridge tops together. *Pinus Sabiniana* is most abundant, however, in the interior of the range, where it occupies extensive areas of rolling savannah in association with *Quercus Douglasii*, and it occasionally occurs in the chaparral.

CUPRESSACEAE

JUNIPERUS CALIFORNICA Carr. (*Sabina californica* Antone.) Canyon slopes and gravelly flood beds, across range, mostly in interior and on east side.

LIBOCEDRUS DECURRENS Torr. Occurs as a planted tree near summit of Copernicus Peak, but is not known to occur naturally within the Mount Hamilton Range. Information relative to the planting of this species was obtained from Dr. W. H. Wright, Director of the Lick Observatory.

CUPRESSUS SARGENTII Jepson var. DUTTONI Jepson, type from Cedar Mountain, *Jepson 7741;* known only from the large colony which occurs on summit of Cedar Mountain.

CUPRESSUS SARGENTII Jepson occurs as a planted tree on Copernicus Peak adjacent to *Libocedrus decurrens* and again near an abandoned homestead about one mile by road down the east side of Copernicus Peak.

TAXACEAE

TORREYA CALIFORNICA Torr. Mount Sizer, *Hendrix 739* (VTM). Known hitherto in the South Coast Ranges only in the Santa Cruz Mountains; typically an outer Coast Range species of moist, shaded localities.

NAJADACEAE

ZANNICHELLIA PALUSTRIS L. Pool, Santa Isabella Valley.

JUNCAGINACEAE

LILAEA SUBULATA Humb. and Bonpl. Drying vernal pools or margins of slowly moving streams.

ALISMACEAE

ALISMA PLANTAGO L. Santa Isabella Creek.

GRAMINEAE

BROMUS RUBENS L. Common across range on dry, open hills.

*BROMUS RIGIDUS Roth. Occasional on dry, open hills, west side of range.

*BROMUS MOLLIS L. (*B. hordaceus* L.) Common across range, on dry, open hills.

*BROMUS ARENARIUS Labill. Common across range on dry, open hills.

BROMUS BREVIARISTATUS Buckl. (*B. subvelutinus* Shear.) Dry slope, west side of Mount Hamilton.

BROMUS CARINATUS Hook. and Arn. Open hills, vicinity Smith Creek.

BROMUS ANOMALUS Rupr. (*B. Porteri* (Coult.) Nash.) West side Mount Hamilton.

PUCCINELLIA SIMPLEX Scribn. Alkaline seepage, Corral Hollow.

FESTUCA OCTOFLORA Walt. Occasional in rocky soil.

FESTUCA PACIFICA Piper. Common across range on open hills.
FESTUCA CONFUSA Piper. Open, dry hills, interior and east side of range.
FESTUCA GRAYI (Abrams) Piper. Occasional across range on open, dry hills.
FESTUCA REFLEXA Buckl. Common across range on open, dry hills.
FESTUCA EASTWOODAE Piper. Occasional across range on openly wooded hills.
FESTUCA MEGALURA Nutt. Occasional across range on open, dry hills.
*FESTUCA DERTONENSIS (All.) Aschers. and Graebn. Occasional across range on dry hills or wooded slopes.
FESTUCA OCCIDENTALIS Hook. and Arn. Copernicus Peak.
FESTUCA ELMERI Scribn. and Merr. Wooded slopes, Mount Hamilton.
*POA ANNUA L. Occasional across range.
POA HOWELLII Vasey and Scribn. Wooded slopes, Mount Hamilton.
*POA PRATENSIS L. West side Mount Hamilton.
POA SCABRELLA (Thurb.) Benth. Common across range on dry, open hills or openly wooded slopes.
POA SECUNDA Presl. (*P. Sandbergii* Vasey.) West side Mount Hamilton.
DISTICHLIS STRICTA (Torr.) Rydb. (*D. spicata* var. *stricta* Scribn.) Alkaline areas, east margin of range.
PHRAGMITES COMMUNIS Trin. Headwaters Arroyo del Puerto.
*LAMARCKIA AUREA (L.) Moench. East of Coyote, *Hendrix 769* (VTM); open hillslope, Oak Ridge, *H. K. Sharsmith 3349* (UC). North of southern California this is a comparatively rare species.
MELICA CALIFORNICA Scribn. (*M. bulbosa* of authors, non Geyer.) Common across range on rocky or openly wooded slopes.
MELICA TORREYANA Scribn. Occasional across range on rocky or wooded slopes.
MELICA IMPERFECTA Trin. var. FLEXUOSA Boland. Open slopes, west side of range.
AGROPYRON SUBSPICATUM (Link) Hitchc. West side Mount Hamilton.
SCRIBNERIA BOLANDERI (Thurb.) Hack. Rocky soil of chaparral slopes, interior of range. Santa Isabella Valley, *H. K. Sharsmith 1856* (UC). This species has been found only in widely separated localities in the South Coast Ranges (Contra Costa County, Mount Hamilton Range, San Luis Obispo County), although it is fairly frequent in the North Coast Ranges and northward to Washington, and it also occurs in the Sierra Nevada. In Mount Hamilton Range it appears to be exclusively a chaparral inhabitant. It is quite common in the vicinity of Arroyo Bayo and Santa Isabella Valley, although an inconspicuous plant.
ELYMUS TRITICOIDES Buckl. Moist soil, Santa Isabella Valley.
ELYMUS CONDENSATUS Presl. Open hills, west side Mount Hamilton.
ELYMUS GLAUCUS Buckl. Stream margins or flood beds, interior of range.
ELYMUS GLAUCUS var. JEPSONI Davy. Tule Lake.
SITANION HANSENI (Scribn.) J. G. Smith. Wooded slopes, east side Mt. Hamilton.
SITANION JUBATUM J. G. Smith. Open or wooded hills or canyon slopes across range.
SITANION HYSTRIX (Nutt.) J. G. Smith. Corrall Hollow.
*HORDEUM MURINUM L. Occasional across range on open hills or exposed flats.
*HORDEUM GUSSONEANUM Parl. West side Mount Hamilton.
HORDEUM JUBATUM L. West side Mount Hamilton.
HORDEUM NODOSUM L. Occasional across range on moist flats near streams.
*LOLIUM MULTIFLORUM Lam. Spring, north side Mount Hamilton.
KOELERIA CRISTATA (Lam.) Pers. Occasional on west side and in interior of range, on dry, rocky, or openly wooded slopes.
*AVENA FATUA L. Common near western base of Mount Hamilton adjacent to cultivated land, uncommon in interior of range.
*AVENA BARBATA Brot. Common across range on open or wooded slopes or flats.

DESCHAMPSIA DANTHONIOIDES (Trin.) Munro. (*Aira danthonioides* Trin.) Occasional on eastern side of Mount Hamilton and in interior of range near streams.
DESCHAMPSIA ELONGATA (Hook.) Munro. West side Mount Hamilton.
*AGROSTIS VERTICILLATA Vill. Stream edge, northeast base Mount Hamilton.

ALOPECURUS HOWELLII Vasey. (*A. saccatus* of authors, non Vasey.) Sag pond, Arroyo Mocho.

*POLYPOGON LUTOSUS (Poir.) Hitchc. Occasional at stream margins, west side and interior of range.

*POLYPOGON MONSPELIENSIS (L.) Desf. Moist areas, north side Mount Hamilton.

*GASTRIDIUM VENTRICOSUM (Gouan) Schinz and Thell. Occasional on west side and interior of range, open ground.

STIPA PULCHRA Hitchc. Occasional across range on openly wooded slopes.

STIPA LEPIDA Hitchc. Oso Creek.

PANICUM CAPILLARE L. var. OCCIDENTALE Rydb. (*P. barbipulvinatum* Nash.) Moist areas, interior and east side of range.

CYPERACEAE

ELEOCHARIS MAMILLATA Lindb. (*E. palustris* (L.) Roem. and Schult. in part.) Occasional in interior of range at creek margins.

ELEOCHARIS ACICULARIS (L.) Roem. and Schult. Occasional in moist places, west side and interior of range.

CAREX PRAEGRACILIS W. Boott. Moist places, interior and east side of range.

CAREX DENSA Bailey. Moist places, west side and interior of range.

CAREX SERRATODENS W. Boott. (*C. bifida* Boott. non Roth.) Occasional across range at creek margins or springs.

CAREX NUDATA W. Boott. Occasional across range at margins of streams.

LEMNACEAE

LEMNA MINOR L. Murietta Springs, Copernicus Peak.

JUNCACEAE

JUNCUS SPHAEROCARPUS Nees. Stream bed, east slope Mount Hamilton.

JUNCUS BALTICUS Willd. (var.?) Colorado Creek, Red Mountains, *C. W. and H. K. Sharsmith 3175* (UC). Immature; possibly var. *montanus* Engelm.

JUNCUS PATENS Meyer. Seepage areas, north side Mount Hamilton.

JUNCUS EFFUSUS L. var. PACIFICUS Fern. and Wieg. (*J. effusus* L. of authors.) Seepage areas, Mount Hamilton.

JUNCUS BUFONIUS L. Moist areas, interior of the range. The variety *congestus* Wahlb. is not recognized here, although the writer does not feel justified in passing judgment on its validity.

JUNCUS OCCIDENTALIS (Coville) Wiegand. (*J. tenuis* Willd. var. *congestus* Engelm.) Moist areas, west side and interior of the range.

JUNCUS XIPHIOIDES Meyer. Moist places, west side of range.

JUNCUS OXYMERIS Engelm. Moist places, east side and base of Mount Hamilton.

JUNCUS ENSIFOLIUS Wikstr. Boggy area, north side Mount Hamilton.

LUZULA MULTIFLORA (Ehrh.) Lej. Wooded areas, west side of range.

LILIACEAE

ZYGADENUS VENENOSUS Wats. Stream margins, east side of range.

ZYGADENUS FREMONTII (Torr.) Torr. Occasional in chaparral across range.

CHLOROGALUM POMERIDIANUM (DC.) Kunth. (*Laothoë pomeridiana* Raf.) Occasional in chaparral, interior of range.

ALLIUM UNIFOLIUM Kell. Stream margins across range; rare. One or two corms arise laterally from short rhizomes which bear the roots.

ALLIUM FIMBRIATUM Wats. Abundant on loose serpentine talus near head of Arroyo del Puerto, Red Mountains: *H. K. Sharsmith 1676, 3143, 3573* (UC); *Elmer 4633* (S). Typically of the desert and mountainous areas of southern California, extending north in the southern Sierra Nevada foothills, and in the inner South Coast Ranges to the Mount Hamilton Range, reappearing in serpentine areas of the North Coast Ranges in Napa and Lake counties.

ALLIUM PARRYI Wats. Abundant on loose serpentine talus near head of Arroyo del Puerto, Red Mountains: *H. K. Sharsmith 1821, 3117* (UC), *Mason* in 1935 (UC). General distribution as for the preceding species; found abundantly in the same area in the Mount Hamilton Range, its northern outpost.

ALIUM AMPLECTENS Torr. Dry stream beds, interior of range.

ALLIUM SERRATUM Wats. Hills and canyon slopes of Red Mountains and east side of range, especially in serpentine rock.

ALLIUM PENINSULARE Lemmon var. CRISPUM (Greene) Jepson. (*A. crispum* Greene.) Talus; Red Mountains, Mount Oso.

ALLIUM BOLANDERI Wats. Wooded slope, east side Mount Hamilton, *H. K. Sharsmith 1171, 1224* (UC); summit Mount Day, *H. K. Sharsmith 3360* (UC). *Allium Bolanderi* was reported from Mount Hamilton by Greene (Erythea 1: 96. 1893). The collections cited above represent the first recollections of the species from Mount Hamilton and confirm Greene's record. Of *A. Bolanderi*, Greene says, "Species not before heard of as from any point south of Humboldt County. We have little more than the oblique corm like bulbs and lateral scapes to judge from; but these in this species are very characteristic." The outer coats of the corms have serrate reticulations, while the more delicate inner coats have obscure, undulate reticulations. One to several of these coated corms may be grouped at the base of the slender scape which arises laterally from the corms. This lateral attachment of scape to corm is, as indicated by Greene, highly distinctive, and it serves to delineate *A. Bolanderi* from other species of superficial resemblance. This diagnostic feature is not brought out in the more recent treatments of California species of *Allium. Allium Bolanderi* appears to be restricted to the North Coast Ranges and Siskiyou Mountains from Siskiyou County and southern Oregon to Lake County, and to the western summit peaks of the Mount Hamilton Range in the South Coast Ranges.

ALLIUM LACUNOSUM Wats. Dry stream beds, infrequent across range.

ALLIUM FALCIFOLIUM Hook. and Arn. (*A. falcifolium* var. *Breweri* Jones; *A. Breweri* Wats.) *Exposed talus*, west side and interior of range. *Allium falcifolium* and *A. Breweri* appear to be identical, but genetic studies are necessary before the question can be adequately decided. Uniting *A. Breweri* with *A. falcifolium*, the older binomial, amplifies the range of the latter species from the Siskiyou Mountains and North Coast Ranges to include as well Mount Diablo, the Mount Hamilton Range, and eastern summit areas of the Santa Cruz Mountains in the South Coast Ranges.

MUILLA SEROTINA Greene. Monument Peak, *Wilson 569* (VTM). The known northern station; a species typically of cismontane southern California and adjacent desert areas; infrequent in inner South Coast Ranges.

BRODIAEA LAXA (Benth.) Wats. (*Tritelia laxa* Benth.) Occasional across range on wooded slopes or in grassland adjacent to streams.

BRODIAEA PEDUNCULARIS (Lindl.) Wats. (*Tritelia peduncularis* Lindl.) Stream margins, Red Mountains. Colorado Creek, *H. K. Sharsmith 3177* (UC); San Antonio Valley, *Mason* in 1935 (UC); Arroyo del Puerto, *Elmer 4348* (S, CA). Not otherwise known from South Coast Ranges, but occurring in moist places in North Coast Ranges from Marin County to Humboldt County.

BRODIAEA HYACINTHINA (Lindl.) Baker. Moist soil, Red Mountains.

BRODIAEA CORONARIA (Salisb.) Jepson. (*Hookera coronaria* Salisb.; *Brodiaea grandiflora* Smith.) Occasional grassy slopes across range, less common than *B. laxa*.

BRODIAEA CAPITATA Benth. (*Dichelostemma capitatum* Wood.) A common species across range on grassy hillsides.

BRODIAEA PULCHELLA Greene. (*B. congesta* Smith; *D. pulchellum* Heller.) West side Mount Hamilton.

CALOCHORTUS LUTEUS Dougl. Common across range on exposed hillsides.

CALOCHORTUS VENUSTUS Dougl. Frequent across range on dry, exposed slopes.

CALOCHORTUS INVENUSTUS Greene. (*C. Nuttallii* of authors, non Torr.; *C. Nuttallii* var. *australis* Munz.) Rocky outcrop above Colorado Creek, Red Mountains, *H. K. Sharsmith 3804* (UC). A common species in southern California, and only rarely found in the South Coast Ranges. The Mount Hamilton Range represents its known northern outpost.

CALOCHORTUS ALBUS (Benth.) Dougl. Occasional on wooded slopes, west side and interior of range.

CALOCHORTUS CLAVATUS Wats. Arroyo del Puerto, *Hoover 3415* (UC). A species found in southern California and frequent to San Luis Obispo County, but occurring only occasionally in South Coast Ranges north of this; not hitherto collected from the Mount Hamilton Range.

CALOCHORTUS UMBELLATUS Wood. On steep canyon slopes beneath chaparral, west side of range. Mainly restricted to North Coast Ranges, the Mount Hamilton Range being its known southern limit.

Fritillaria agrestis Greene. (*F. succulenta* Elmer.) Tesla hills above Corral Hollow.

Fritillaria lanceolata Pursh. (*F. lanceolata* var. *floribunda* Benth.) Occasional across range on wooded slopes.

Fritillaria folcata (Jepson) Beetle, Madroño 7:148. 1944. (*F. atropurpurea* var. *folcata* Jepson.) A species restricted to serpentine talus of inner South Coast Ranges. Known only from the type ("San Benito Co., San Benito Peak," *Jepson 2715*), and from the Red Mountains of Mount Hamilton Range: Adobe Creek *H. K. Sharsmith 1671* (UC), flower and *3579* (UC), fruit; Colorado Creek, *Beetle 17* (UC), *Carter 1047* (UC).

Disporum Hookeri (Torr.) Britt. Infrequent on wooded slopes, Mount Hamilton. In the South Coast Ranges typically occurring only near the coast. The colony at north base of Mount Hamilton (*H. K. Sharsmith 1209*, UC) appeared to be persisting vegetatively; in 1934 only abortive flowers were found, and in 1935 none were found.

Smilacina sessilifolia (Baker) Nutt. (*Vagnera sessilifolia* Greene.) Shady areas, west side of range; infrequent. In the South Coast Ranges a typically outer Coast Range species.

Smilacina amplexicaulis Nutt. (*Vagnera amplexicaulis* Greene.) Dense shade, west side of range; infrequent. Mainly limited to outer Coast Ranges.

Trillium sessile L. var. giganteum Hook. and Arn. (*T. chloropetalum* Howell.) Wooded slopes, west side of range; infrequent. In central California mainly limited to outer Coast Ranges.

IRIDACEAE

Iris longipetala Herbert. Boggy area, west side Mount Hamilton.

Iris macrosiphon Torr. Wooded areas, west side of range.

Sisyrinchium bellum Wats. Occasional on wooded hillslopes across range.

ORCHIDACEAE

Habenaria unalaschensis (Spreng.) Wats. (*Piperia unalaschensis* Rydb.) Wooded slope, west side Mount Hamilton; infrequent.

Habenaria Michaeli Greene. Rocky ridge, Arroyo del Puerto.

Epipactis gigantea Dougl. (*Serapias gigantea* A. A. Eaton.) North base Mount Hamilton; infrequent.

SALICACEAE

Populus Fremontii Wats. Occasional to abundant, stream beds of interior and north and east sides of range.

Populus trichocarpa Torr. and Gray. Madrone Springs, *Hendrix 722* (VTM).

Salix laevigata Bebb. Along streams, interior and east side of range.

Salix laevigata Bebb. var. araquipa (Jepson) Ball. Arroyo Mocho.

Salix Hindsiana Benth. Burnt Hills, stream margins.

Salix melanopsis Nutt. Santa Isabella Creek, east base Mount Hamilton.

Salix lasiolepis Benth. Stream margins, interior of range.

Salix Breweri Bebb. Abundant along streams in isolated localities, Red Mountains. A distinctive species restricted to inner Coast Ranges from San Luis Obispo to Colusa counties; in the Mount Hamilton Range occurring only in areas of serpentine rock.

BETULACEAE

Alnus rhombifolia Nutt. Occasional to abundant along streams of west side of range. There is a single large tree at Aquarius Springs; Greene (Erythea 1: 95. 1893) lists this as *Alnus rubra* Bong., apparently an inadvertence.

FAGACEAE

Quercus lobata Nee. Common across range on rolling hills or valley flats.

Quercus Garryana Dougl. Near Old Gilroy, *Hendrix 728* (VTM); Mount Hamilton, *Eastwood 12432* (CA). Not otherwise known from Mount Hamilton Range, although more frequent in Santa Cruz Mountains. Hendrix' specimen was taken from a sizeable grove of trees which were very distinct in appearance from Q. *morehus* and other oaks of this area.

QUERCUS DOUGLASII Hook. and Arn. Common across range on rolling hills below 3500 feet, usually associated with *Pinus Sabiniana*.

QUERCUS DUMOSA Nutt. Locally abundant across range as an occasional chaparral element.

QUERCUS DURATA Jepson. Occasional chaparral element across range, apparently confined to areas of serpentine rock, common in Red Mountains.

QUERCUS CHRYSOLEPIS Liebm. West side of range at altitudes above 2250 feet, most abundant on north slopes near summit ridges; also on higher peaks of interior of range; often associated with *Q. Wislizenii*.

QUERCUS AGRIFOLIA Nee. Occasional on Mount Hamilton mainly on west side.

QUERCUS WISLIZENII DC. Occasional across range as a chaparral constituent; abundant on Mount Hamilton and west side of range as an associate of *Q. chrysolepis* on north facing slopes.

QUERCUS KELLOGGII Newb. West side and interior of range above 2000 feet.

QUERCUS MOREHUS Kell. Near Lost Lake, *Hendrix 721* (VTM). A hybrid of *Q. Wislizenii* x *Q. Kelloggii*, to be expected in regions where both parent species occur.

JUGLANDACEAE

JUGLANS HINDSII (Jepson) Jepson. (*J. californica* Wats. var. *Hindsii* Jepson.) North end Adobe Valley, *H. K. Sharsmith 3567a* (UC). A central Californian species occurring as a native only in a few isolated, highly localized areas; not reported hitherto from the Mount Hamilton Range. The Adobe Valley locality is far removed from any habitation, and the single tree appears not to have been planted by white man.

URTICACEAE

URTICA GRACILIS Ait. var. HOLOSERICEA (Nutt.) Jepson. (*U. holosericea* Nutt.) Occasional near springs, west side and interior.

*URTICA URENS L. Stream bed, Arroyo del Puerto.

LORANTHACEAE

PHORADENDRON FLAVESCENS Nutt. var. MACROPHYLLUM Engelm. (*P. longispicum* Trel.) Common parasite on *Populus Fremontii*, east and north sides of range.

PHORADENDRON VILLOSUM Nutt. (*P. flavescens* var. *villosum* Engelm.) Parasite on *Quercus*, west side Mount Hamilton.

ARCEUTHOBIUM CAMPYLOPODUM Engelm. (*Razoumofskya campylopoda* Kuntze.) Parasite on *Pinus Sabiniana*.

POLYGONACEAE

POLYGONUM PARRYI Greene. Rocky, dry sag pond in chapparal near Santa Isabella Valley, *H. K. Sharsmith 3311* (UC). The only known record of this species in the South Coast Ranges. It occurs in the higher parts of the North Coast Ranges northward to Washington, and is also reported from the central Sierra Nevada and the Cuyamaca Mountains (San Diego County).

*RUMEX CRISPUS L. Occasional weed in valley flats.

*RUMEX CONGLOMERATUS Nutt. Occasional weed in moist, valley flats.

*RUMEX ACETOSELLA L. Grassy slope, Mount Hamilton.

RUMEX SALICIFOLIUS Weinm. Occasional in moist flats across range.

PTEROSTEGIA DRYMARIOIDES Fisch. and Mey. Moist wooded slope, Arroyo del Puerto.

CHORIZANTHE MEMBRANACEA Benth. (*Eriogonella membranacea* Goodm.) Common vernal species on dry, open slopes across range.

CHORIZANTHE UNIARISTATA Torr. and Gray. Red Mountains.

CHORIZANTHE CLEVELANDII Parry. Occasional in chaparral or dry, sandy flood beds of streams. Santa Isabella Creek, *H. K. Sharsmith 1132* (UC); between Arroyo Bayo and San Antonio Valley, *H. K. Sharsmith 3298* (UC). A species of intermittent distribution, occurring in the inner Coast Ranges from Mendocino County to Tulare County, and apparently not found between Lake County and the Mount Hamilton Range. The above are the first collections made between the San Carlos Range and Lake County. *Chorizanthe Clevelandii* is easily mistaken for *C. uniaristata*, with which it is closely allied, but from which it differs in having uncinate, nearly equal involucral spines, and a perianth of different size and shape.

CHORIZANTHE POLYGONOIDES Torr. and Gray. (*Acanthogonum polygonoides* Goodm.) Talus, Arroyo del Puerto.

CHORIZANTHE PERFOLIATA Gray. (*Mucronea perfoliata* Heller.) Arroyo del Puerto, *H. K. Sharsmith 1755* (UC); Puerto Canyon, *Brewer 1261* (UC). The above citations represent the known northern limit of this typically desert species.

ERIOGONUM ANGULOSUM Benth. Grassy hillslopes and steep talus, east side of range. A southern Californian species which reaches its northern limit in eastern Contra Costa County just north of the Mount Hamilton Range.

ERIOGONUM INERME (Wats.) Jepson. (*E. vagans* Wats.) Occasional in rocky soil of chaparral or in dry, sandy flood beds of streams. Santa Isabella Peak, *H. K. Sharsmith 3382* (UC); Red Mountains, *H. K. Sharsmith 3792* (UC). An uncommon species occurring from southern California northward in the inner South Coast Ranges to Mount Hamilton Range; not previously reported north of San Carlos Range (San Benito County).

ERIOGONUM VIRGATUM Benth. (*E. vimineum* subsp. *virgatum* Stokes.) A common autumnal species of open hills and valley flats; across range, but most abundant in interior.

ERIOGONUM VIMINEUM Dougl. A common late spring species across range, but most abundant in interior on open, rocky slopes, often occurring in chaparral.

ERIOGONUM COVILLEANUM Eastw. Proc. Calif. Acad. Sci. ser. 4, **20**:138. 1931; type, "road from the summit of Mount Hamilton to Livermore" (Arroyo Bayo), *Eastwood*, April 26, 1927 (CA!). Shale and serpentine talus, interior and east side of range. Arroyo Bayo. *H. K. Sharsmith 3061* (UC), topotype; Colorado Creek, *H. K. Sharsmith 3162* (UC); Adobe Creek, *H. K. Sharsmith 3576* (UC); Arroyo del Puerto, *H. K. Sharsmith 3122a* (UC). A species which is restricted to the inner South Coast Ranges from Santa Clara County to Monterey County. It is related to the *E. vimineum* complex, but in addition to its distinctive morphological features it blooms in April and May whereas *E. vimineum* and allies bloom in August and September.

ERIOGONUM SAXATILE Wats. Rock crevices, northeast ridge of Copernicus Peak near summit of Mount Hamilton, *H. K. Sharsmith 1299* (UC), *Carter 648* (UC). *Eriogonum saxatile* was collected long ago near the summit of Mount Hamilton by E. L. Greene (Erythea **1**:84. 1893)—"probably the northern limit of the species." The above citations represent the first recollections of this species at Mount Hamilton, the known northern outpost of the species. Despite Greene's early collections and several more recent collections from the eastern side of the Santa Cruz Mountains and one from the Pinnacles of San Benito County (Howell, Leafl. West. Bot. **2**:99. 1938), it has not been commonly realized that this species occurs in the South Coast Ranges.

ERIOGONUM WRIGHTII Torr. (*E. trachygonum* Torr.; *E. trachygonum* subsp. *Wrightii* Stokes.) Occasional to abundant across range on rocky ridges.

ERIOGONUM FASCICULATUM Benth. var. FOLIOLOSUM (Nutt.) Stokes. Rocky canyon slopes, usually a chaparral element, east side of range. This variety represents the most abundant phase of the species. It is common in cismontane southern California, and occurs northward in the inner South Coast Ranges to Mount Hamilton Range. Corral Hollow is its northern outpost.

ERIOGONUM NUDUM Dougl. var. AURICULATUM (Benth.) Tracy. Dry, rocky slopes, Mount Hamilton.

ERIOGONUM NUDUM Dougl. var.? Occasional on rocky slopes or in chaparral, interior and east side of range. Seeboy Ridge, *H. K. Sharsmith 3879* (UC); Adobe Creek, *H. K. Sharsmith 3909* (UC); Arroyo del Puerto, *Hoover 2629* (UC). These specimens approach the variety *pubiflorum* Benth. of southern California. The involucres are one to three in a place, and the calyx is deep yellow. In the variety *pubiflorum*, however, the calyx is hairy toward the base, and in the Mount Hamilton Range plants it is glabrous.

ERIOGONUM UMBELLATUM Torr. var. STELLATUM Jones. (*E. stellatum* Benth.; *E. umbellatum* subsp. *stellatum* Stokes; *E. umbellatum* var. *bahiaeforme* Jepson; *E. trichotomum* Small (Bull. Torr. Bot. Club **25**:43. 1898), type from Mount Hamilton, Greene). North slope Mount Hamilton near summit, *Stokes 100* (UC); listed by

Greene (*E. stellatum*, Erythea **1**:83. 1893) as "very common on sunny slopes. ... near the summit of Mount Hamilton."

CHENOPODIACEAE

*CHENOPODIUM ALBUM L. Occasional in flood beds of streams, west side of range.
CHENOPODIUM CALIFORNICUM Wats. Occasional on wooded slopes, west side and interior of range.
ATRIPLEX SERENANA Nels. (*A. bracteosa* Wats.) Adobe Creek.
*SALSOLA KALI L. var. TENUIFOLIA G. F. W. Mey. Lick Observatory. summit Mount Hamilton.

AMARANTHACEAE

*AMARANTHUS BLITOIDES Wats. Flood bed Santa Isabella Creek.
AMARANTHUS CALIFORNICUS (Moq.) Wats. Mount Day Ridge.

AIZOACEAE

*GLINUS LOTOIDES L. Abundant in dried mud, edge of sag pond, Mount Day Ridge, *Carter 1201* (UC), *H. K. Sharsmith 3837* (UC). An adventive species reported in California at several isolated stations. *Carter 1201* is the first collection from Mount Hamilton Range.

PORTULACACEAE

CALYPTRIDIUM MONANDRUM Nutt. Abundant in chaparral, divide between Arroyo Bayo and San Antonio Valley, *H. K. Sharsmith 3077* (UC); Corral Hollow, *Hoover 3044* (UC). Typically a species of southern California, Colorado and Mohave deserts, and western Arizona; occurring infrequently in the inner South Coast Ranges as far north as Mount Hamilton Range, known there only from these two collections.
CALYPTRIDIUM PARRYI Gray. Chaparral, northeast side Mount Santa Isabella, *H. K. Sharsmith 3381* (UC). A species of the mountainous areas of southern California, not heretofore reported north of Mount Pinos in the Tehachapi Mountains. Determined by R. S. Ferris who reports, "Somewhat aberrant form (slender), but well within the variation of the species."
CALANDRINIA CILIATA (Ruiz and Pavon) DC. var. MENZIESII (Hook.) Macbr. (*C. caulescens* H. B. K. var. *Menziesii* Gray.) Common vernal annual in moist valley flats across range.
MONTIA FONTANA L. Moist areas, interior of range.
CLAYTONIA PERFOLIATA Donn. (*Montia perfoliata* Howell; *Limnia perfoliata* How.; *Claytonia nubigena* Greene; *M. perfoliata* var. *nubigena* Jepson; *Limnia nubigena* Heller; *Montia perfoliata* var. *depressa* (Gray) Jepson; *M. perfoliata* var. *parviflora* (Dougl.) Jepson.) Shady wooded slopes or on steep talus in chaparral, across range.
CLAYTONIA GYPSOPHILOIDES Fisch. and Mey. (*Montia gypsophiloides* Howell; *Limnia gypsophiloides* Heller; *L. diaboli* Rydb.) Across range, but most abundant on rocky slopes which occur as openings in chaparral or on steep talus. Distinctions between *Limnia gypsophiloides* and *L. diaboli* Rydb., as outlined by Rydberg (North American Flora **21**:311. 1932), tenuous at best, do not hold in Mount Hamilton Range material.
LEWISIA REDIVIVA Pursh. Rocky ridges or slopes, interior of range, rare.

CARYOPHYLLACEAE

*CERASTIUM VISCOSUM L. Occasional in grasslands, across range.
CERASTIUM ARVENSE L. var. MAXIMUM Holl. and Britt. (*C. viride* Heller, Muhl. **2**:281. 1907. Type from Alum Rock Park, Mount Hamilton Range, *Heller 8485*.) Alum Rock Park.
*STELLARIA MEDIA (L.) Cyr. Uncommon in Mount Hamilton Range except in cultivated area near base of western slope.
STELLARIA NITENS Nutt. Occasional across range on grassy, openly wooded slopes.
SAGINA OCCIDENTALIS Wats. Santa Isabella Creek.
*SAGINA APETALA Ard. var. BARBATA Fenzl. Occasional in moist places, interior and east side of range.
ARENARIA MACROPHYLLA Hook. (*Moehringia macrophylla* Torr.) Common on wooded slopes near summits of higher peaks on west side of range.
ARENARIA DOUGLASII Fenzl. Frequent across range on rocky slopes.

Arenaria Douglasii Fenzl. var. **emarginata** var. nov. Petalis obovatis plerumque emarginatis nonnunquam irregulariter dentatis vel nihil nisi obtusis 2-3 mm. longis sepala in fructu aequantibus vel vix superantibus, staminibus aequalibus 2 mm. longis brevioribus quam petalis sepalisque, antheris 0.2-0.3 mm. longis.

Connate bases of leaves not white-scarious, lower leaves up to 6 cm. long, withering early; flowers 3 mm. long, 2.5 mm. wide; sepals 2-2.5 mm. long in flower, 3-3.5 mm. long in fruit, red tipped or sometimes red throughout; petals obovate, usually emarginate, occasionally merely obtuse, 2-3 mm. long, often slightly longer than sepals before anthesis, about equal to fruiting sepals after anthesis; stamens equal, 2 mm. long, shorter than petals and sepals, bidentate gland small, anthers 0.2-0.3 mm. long; styles feathery to base.

Type. West talus of canyon, Adobe Creek, Stanislaus County, Red Mountains, Mount Hamilton Range of South Coast Ranges, California, altitude 1800 feet, April 22, 1936, *H. K. Sharsmith 3575* (UC). Other specimens. Adobe Creek, *H. K. Sharsmith 1670* (topotype, immature, UC); Adobe Creek, *H. K. Sharsmith 3585* (UC), Carter *1158* (UC), Arroyo del Puerto, *H. K. Sharsmith 3762* (UC).

The close affinity of *Arenaria Douglasii* and *A. Douglasii* var. *emarginata* is indicated by the general features of the plants, and also by the presence of yellowish bidentate glands on those stamens which are opposite the sepals. The other two annual Californian species of the section *Alsine* (*A. californica* and *A. pusilla*) do not possess these glands.

The most distinctive features of *Arenaria Douglasii* var. *emarginata* lie in the flowers, and particularly in the petals and stamens, although intergradation with the species occurs to some extent. Thus, in the variety the petals may be irregularly toothed or merely obtuse, but the emarginate condition prevails, and the petals do not or scarcely exceed the fruiting sepals; the stamens are 2 mm. long, shorter than petals and sepals, and equal; the anthers are 0.2-0.3 mm. long, and the bidentate gland at the base of the five stamens is relatively small. In the species the petals are always obtuse, and are often twice as long as the fruiting sepals, although occasionally they may be as short as the sepals; the stamens are up to 4 mm. long, often exceeding the sepals and often unequal; the anthers are 0.5-0.7 mm. long; and the bidentate gland is relatively large.

Arenaria Douglasii var. *emarginata* (*H.K.S. 3956*) and *A. Douglasii* (*H.K.S. 3957*) were grown from seed, and observed simultaneously from germination to maturity. The differences referred to in the preceding paragraph were observable in the cultivated plants.

Arenaria Douglasii is found with *A. Douglasii* var. *emarginata* in the Red Mountains of the Mount Hamilton Range, but their colonies do not intermingle. *Arenaria Douglasii* is common on exposed, dry rock outcroppings throughout the range, but so far *A. Douglasii* var. *emarginata* has been found only on the unstable serpentine or shale talus of the Red Mountains.

ARENARIA CALIFORNICA (Gray) Brewer. Occasional in interior of range on rocky chaparral slopes or open hillsides. A rare species in the South Coast Ranges.

ARENARIA PUSILLA Wats. Frequent in rocky soil of chaparral, interior of range. Sulphur Spring Creek, *C. W. and H. K. Sharsmith 3435* (UC); Arroyo Mocho near Colorado Creek, *H. K. Sharsmith 3514* (UC); Sugarloaf Mountain, *H. K. Sharsmith 3635* (UC); Arroyo Bayo, *H. K. Sharsmith 1982a* (UC). For many years *A. pusilla* was known in Washington and Oregon, and as far south as Humboldt County in the North Coast Ranges of California. A citation by Munz (Man. S. Calif. Bot. 163. 1935; Laguna Mountains, San Diego, *Munz 9672*, P!) extends our knowledge of the range of this species to the southern extremity of California. The citations above are the first record of specimens collected between Humboldt County and San Diego County, and they partly eliminate the discontinuity in the known range of the species. *Arenaria pusilla* probably occupies other favorable localities in the Coast Ranges, but its small size prevents easy detection.

The Mount Hamilton Range specimens collected in 1936 (*3435, 3514, 3635*) are uniformly smaller than available herbarium material of *A. pusilla*, and are also smaller than Watson's type sheet (G!) and Munz' San Diego specimen, but they are similar in all other respects. Their depauperate nature seems to have been conditioned by aridity, as 1936 was a year of low rainfall in this region. The collection made in 1935 *(1982a),* a year of average rainfall for the region, shows normal sized plants.

Arenaria pusilla belongs to the section *Alsine* Benth. and Hook. Its closest relative is *A. californica* (Gray) Brewer. Both species occur in the Mount Hamilton Range. They are distinguished by features which seem not to intergrade. *Arenaria pusilla* is smaller; the sepals are narrowly lanceolate and acuminate; the petals are shorter than the sepals or wanting; and the stamens are two-thirds the length of the sepals. *Arenaria californica* is larger; the sepals are oblong-ovate and acute; the petals are one-half again the length of the sepals; and the stamens are as long as the sepals.

SPERGULARIA ATROSPERMA R. P. Rossbach. Alkaline seepage, Elk Ravine, Corral Hollow, *Ferris 9412* (S).

SPERGULARIA SALIGNA J. and C. Presl. Corral Hollow.

LOEFLINGIA SQUARROSA Nutt. Rocky flood plain, Corral Hollow, *Carter 788* (UC). An infrequently collected species.

*HERNIARIA CINEREA DC. Grassland, east side of range.

*SILENE GALLICA L. Occasional on grassy slopes, west side of range.

RANUNCULACEAE

PAEONIA BROWNII Dougl. Large colony persisting mainly by vegetative propagation, buds small and abortive, flowers few or none; east side Mount Hamilton, *H. K. Sharsmith 3947* (UC). This colony was located by G. L. Stebbins, Jr.; it is discussed in his recent paper on *Paeonia* (Madroño 4:252-260. 1938).

ISOPYRUM OCCIDENTALE Hook. and Arn. Occasional on north facing, wooded slopes, west side and interior of range; frequent on east side of Mount Hamilton in moist, loose humus under *Quercus Kelloggii. Isopyrum occidentale* is a highly localized and infrequently collected although widely distributed species. Isolated stations occur in Butte, Amador, Mariposa, and Kern counties in the Sierra Nevada; Vaca Mountains, Mount Hamilton Range, Santa Cruz Mountains, and San Carlos Range in the Coast Ranges; Tehachapi Mountains in southern California.

ISOPYRUM STIPITATUM Gray. Moist, north facing, wooded slopes, west side and interior of range, usually growing with *I. occidentale;* rare. *Isopyrum stipitatum* is, like *I. occidentale,* a highly localized species, but it is considerably less well known. The reported stations for it occur mainly in Modoc and Siskiyou counties. Comparison of the two species from the Mount Hamilton Range, where they grow together intimately, leaves no doubt as to their specific differentiation. Careful observation shows several definite field characters by which the two species may be distinguished: *I. stipitatum* is about two-thirds the size of *I. occidentale,* and its fascicled roots are truncate, whereas in *I. occidentale* they are long attenuate; the smaller leaves of *I. stipitatum* are pale and glaucous above as well as below, and the ultimate leaf segments are much narrower, more completely dissected, and less fan-shaped than in *I. occidentale.* The follicles of *I. stipitatum* are on peduncles which are shorter than the leaves and strongly reflexed, so that the fruits lie on or near the surface of the soil; in *I. occidentale* the peduncles are longer than the leaves and strictly erect, bringing the follicles well above the foliage. It is possible that the apparent rarity of *I. stipitatum* may be due partly to its inconspicuous habit and to the ease with which it could be passed over for small-sized plants of the more obvious *I. occidentale* where the two species grow together.

AQUILEGIA FORMOSA Fisch. subsp. TRUNCATA (Fisch. and Mey.) Jones. (*A. truncata* Fisch. and Mey.) Wooded slopes, west side and interior of range.

AQUILEGIA TRACYI Jepson. Santa Isabella Creek, Mount Hamilton, *H. K. Sharsmith 1215* (UC); margin of spring, Red Mountains, *H. K. Sharsmith 3895* (UC). It has not been recorded hitherto from the Mount Hamilton Range.

DELPHINIUM NUDICAULE Torr. and Gray. Occasional on rocky slopes across range.

DELPHINIUM PATENS Benth. (*D. decorum* of authors, non Fisch. and Mey.) Frequent on wooded, north facing slopes, across range.

DELPHINIUM PATENS Benth. x D. NUDICAULE Torr. and Gray. East side Mount Hamilton, *H. K. Sharsmith 3052,* flowers, *3224,* seeds (UC). On the rocky chaparral clearing where the specimens were found, *D. nudicaule* and *D. patens* were common in separate colonies. At the contact zcne where individuals of the species of one *Delphinium* colony overlapped with individuals of the other colony, several plants were found which could be interpreted only as natural hybrids between *D. patens* and *D. nudicaule*. The rich purple-red corolla color, unfortunately somewhat faded with pressing, is the most outstanding indication of this hybrid ancestry. Except for color, the flower is closer to that of *D. patens*, but the glandular-pubescent follicles are like those of *D. nudicaule* instead of glabrous as in *D. patens*. The large, glabrous, succulent and mostly basal leaves with broad segments, and the woody, fibrous root system closely approach *D. nudicaule*. Such a hybrid involves a cross between two sections of the genus, Section *Phoenicodelphis*, to which belongs *D. nudicaule*, and Section *Delphiniastrum*, to which belongs *D. patens*.

DELPHINIUM CALIFORNICUM Torr. and Gray. Dry ravine near San Antonio Creek.

DELPHINIUM CALIFORNICUM Torr. and Gray var. INTERIUS Eastw. Leafl. West. Bot. 2:137. 1938, type from Hospital Canyon, Mount Hamilton Range, *Eastwood* and *Howell 5796* (CA). Interior and east side of range, infrequent in dry ravines. This variety is possibly distinct from *D. californicum*, from which it may be separated by its glabrous or glabrate flowers and persistent leaves. It appears to be limited in distribution to the Mount Hamilton Range and Mount Diablo.

DELPHINIUM VARIEGATUM Torr. and Gray. Common late spring species on openly wooded hills and in valleys, interior and east side of range. The typical phase of the species is not found in central California, and the Mount Hamilton Range collections should be referred to a variety or subspecies.

DELPHINIUM HESPERIUM (Brew. and Wats.) Gray. Common late spring species in openly wooded hills and valleys, mostly in interior of range.

DELPHINIUM HESPERIUM Gray var. SEDITIOSUM Jepson. Openly wooded hills, interior and east side of range.

DELPHINIUM PARRYI Gray. Occasional, interior of range on openly wooded hills. Common in southern California, but here at the northern margin of its distribution.

DELPHINIUM PARRYI Gray x D. VARIEGATUM Torr. and Gray? San Antonio Valley, *H. K. Sharsmith 3096b* (UC). A small colony of this apparent hybrid occurred under oaks near *D. variegatum* and *D. Parryi*.

THALICTRUM POLYCARPUM (Torr.) Wats. Wooded slopes across range.

MYOSURUS LEPTURUS (Gray) Howell. (*M. aristatus* Benth. var. *lepturus* Jepson; *M. apetalus* var. *lepturus* Gray.) Sag pond between Arroyo Bayo and Santa Isabella Valley.

MYOSUROS MINIMUS L. Moist depressions or in chaparral, east side and interior of range.

MYOSUROS MINIMUS L. var. FILIFORMIS Greene. Vernal pool, Arroyo Mocho. Questionably distinct from the species.

RANUNCULUS CALIFORNICUS Benth. Common vernal species across range on moist slopes or valley flats.

RANUNCULUS HEBECARPUS Hook. and Arn. Occasional on wooded slopes across range.

RANUNCULUS TRICHOPHYLLUS Chaix. (*R. aquatilis* L. var. *trichophyllus* Gray; *R. aquatilis* L. var. *capillaceus* DC.) Vernal pools or sluggish streams, interior of range.

RANUNCULUS TRICHOPHYLLUS Chaix. var. HISPIDULUS (E. Drew) W. Drew. (*R. aquatilis* of authors, non L.) Sag pond, Mount Day Ridge.

CLEMATIS LASIANTHA Nutt. Occasional on brush or chaparral slopes, across range.

BERBERIDACEAE

BERBERIS DICTYOTA Jepson. (*B. californica* Jepson; *Odostemon dictyota* Abrams; *Berberis aquifolium* var. *dictyota* Jepson.) Brushy slopes, interior of range; infrequent.

BERBERIS PINNATA Lag. Smith Creek.

LAURACEAE

UMBELLULARIA CALIFORNICA Nutt. Frequent on west side of range, wooded or brushy slopes, infrequent in interior of range.

PAPAVERACEAE

PLATYSTEMON CALIFORNICUS Benth. Frequent vernal annual, open slopes or valley flats across range.

ARGEMONE PLATYCERAS Link and Otto. (*A. platyceras* var. *hispida* (Gray) Prain.) Dry stream bed, Arroyo del Puerto.

PAPAVER HETEROPHYLLUM (Benth.) Greene. Occasional vernal annual across range on wooded slopes.

ESCHSCHOLTZIA CALIFORNICA Cham. (*E. crocea* Benth.) Common vernal annual across range, locally abundant in colonies on grassy hillslopes.

ESCHSCHOLTZIA CAESPITOSA var. HYPECOIDES (Benth.) Gray. Occasional on grassy slopes, interior and east side of range. Arroyo del Puerto, *H. K. Sharsmith 1740, 1810* (UC), *Hoover 3361* (UC); between Arroyo Bayo and San Antonio Valley, *H. K. Sharsmith 3076* (UC). These collections are dubiously referred to this variety. The white, scale-like hairs and finer pubescence, particularly on *H. K. S. 1740* and *Hoover 3361*, suggest *E. Lemmonii*.

ESCHSCHOLTZIA CAESPITOSA var. RHOMBIPETALA (Greene) Jepson. Infrequent on grassy slopes or talus, interior and east side of range.

FUMARIACEAE

DICENTRA CHRYSANTHA (Hook. and Arn.) Walp. Abundant in isolated colonies in interior and east side of range; often on burned-over land.

CRUCIFERAE

THELYPODIUM FLAVESCENS (Hook.) Wats. (*Caulanthus flavescens* Payson; *Streptanthus Dudleyi* Eastw.) Corral Hollow.

THELYPODIUM LASIOPHYLLUM (Hook. and Arn.) Greene. (*Caulanthus lasiophyllus* Payson; *Guillenia lasiophylla* Greene.) Occasional vernal annual across range on grassy slopes.

STREPTANTHUS COULTERI (Wats.) Gray var. LEMMONII (Wats.) Jepson. (*S. Parryi* Greene; *S. Lemmonii* Jepson; *Caulanthus Lemmonii* Wats.) Occasional along east base of range.

STREPTANTHUS BREWERI Gray. Unstable talus, Red Mountains, the type locality. Summit of mountain north of Camp 75 (near head of Arroyo del Puerto), *Brewer 1268* (isotype, UC!); head of Arroyo del Puerto, *C. W. and H. K. Sharsmith 3149* (UC); Colorado Creek canyon, *C. W. and H. K. Sharsmith 3167* (UC); Red Mountain, *Elmer 4345* (UC). A distinctive species restricted to rocky, serpentine slopes mainly of unstable talus, occurring in the Lake County area of inner North Coast Ranges, and in Mount Hamilton Range and San Carlos Range of inner South Coast Ranges.

STREPTANTHUS GLANDULOSUS Hook. (*S. Mildredae* Greene, type from Mount Hamilton, *Mildren Holden; Euclisia Mildredae* Greene.) Frequent across range on rocky slopes. The Mount Hamilton plants distinguished as *S. Mildredae* by E. L. Greene do not warrant separation from the polymorphic *S. glandulosus*.

STREPTANTHUS ALBIDUS Greene. (*S. glandulosus* Hook. var. *albidus* Jepson.) Serpentine outcrop, Metcalfe road, west side Mount Hamilton near base, *H. K. Sharsmith 3956* (UC). A little known species limited to the west base of Mount Hamilton Range and to Mount Diablo; restricted to serpentine. The type was collected in the Mount Hamilton Range four miles south of San Jose above Cincas Creek by Rattan in 1886 (isotype, S!). A subspecies of *S. glandulosus* according to J. L. Morrison (A monograph of the section *Euclisia* Nutt., of *Streptanthus* Nutt. Thesis, U. of Calif., 1941. 103 mss. pp.).

STREPTANTHUS CALLISTUS Morrison, Madroño 4: 205. 1938. Shale talus, Arroyo Bayo, Mount Hamilton Range, May 5, 1935, *C. W. and H. K. Sharsmith 3074* (type, UC!). A very narrow endemic, found only at the type locality; topotypes collected by Keck and Clausen in 1937 ("very rare"), and by Morrison in 1937 ("mature siliques and seeds") and 1938 ("abundant in the area").

STREPTANTHUS LILACINUS Hoover, Leafl. West. Bot. 1:226. 1936. Corrall Hollow, Mount Hamilton Range, April 7, 1935, *Eastwood and Howell 2111*, type (isotype, CA!); Arroyo del Puerto, *H. K. Sharsmith 1592* (UC).

DESCURAINIA PINNATA (Walt.) Britt. subsp. MENZIESII (DC.) Detling. (*Sisymbrium pinnatum* Greene.) Occasional on grassy slopes of interior and east side of range.

*BRASSICA CAMPESTRIS L. Abundant in agricultural lands at west base Mount Hamilton, but uncommon elsewhere across range.

*BRASSICA ARVENSIS (L.) Rabenh. Uncommon.

*BRASSICA INCANA (L.) Meigen. (*B. adpressa* Boiss.) Uncommon.

BARBAREA ORTHOCERAS Ledeb. var. DOLICHOCARPA Fern. (*B. vulgaris* of authors in part, non R. Br.) Occasional in moist areas, west side and interior of range.

RADICULA NASTURTIUM-AQUATICUM (L.) Brit. and Rendle. (*Rorippa Nasturtium-aquaticum* Schinz. and Thell.) North side Mount Hamilton; bog.

RADICULA CURVISILIQUA (Hook.) Greene. (*Rorippa curvisiliqua* Bessey.) Occasional in moist places, west side and interior of range.

CARDAMINE OLIGOSPERMA Nutt. Wooded slopes, west base of range.

DENTARIA INTEGRIFOLIA Nutt. var. CALIFORNICA (Nutt.) Jepson. (*Cardamine californica* Greene.) Common on wooded, north-facing slopes across range.

ARABIS GLABRA (L.) Bernh. Grassy slope, Colorado Creek.

ARABIS BREWERI Wats. var. TYPICA Rollins. Rocky outcrops, 2000 to 4000 feet, west side and interior of range. In the present survey, *A. Breweri* was found to be rare near the summit of Mount Hamilton, but references to this species by Greene (Erythea **1**: 87. 1893) and Heller (Muhl. **2**:285. 1907) imply that it was once more abundant there.

ERYSIMUM CAPITATUM Greene. (*E. asperum* of authors.) Occasional on dry, rocky slopes across range.

TROPIDIOCARPUM GRACILE Hook. Occasional on grassy slopes across the range.

TROPIDIOCARPUM CAPPARIDEUM Greene. Mountain House (Altamont Pass), *Rose 33019* (UC); between Altamont Pass and Patterson Pass, *Mason 6818* (UC). Also at San Joaquin County entrance to Corral Hollow according to H. L. Mason (oral communication). An endemic restricted to the area lying between Byron Hot Springs and the northeastern borders of the Mount Hamilton Range.

LEPIDIUM NITIDUM Nutt. Common vernal annual across range on grassy slopes.

LEPIDIUM LATIPES Hook. Summit of Red Mountains.

PLATYSPERMUM SCAPIGERUM Hook. Headwaters of Alameda Creek north of Packard Ridge, *Mason 7209* (UC), *Ferris and Bacigalupi 8285* (S), the only records for this species from the Coast Ranges. It occurs from the northern Sierra Nevada north to Washington and Idaho, reaching its highest developments in the northwest.

*CAPSELLA BURSA-PASTORIS (L.) Medic. (*Bursa pastoris* Dorsten.) Established across range at isolated localities.

CAPSELLA PROCUMBENS (L.) Fries. (*Hutchinsia procumbens* Desv.) Grassy areas, east side of range.

DRABA UNILATERALIS Jones. (*Athysanus unilateralis* Jepson.) Occasional, east side of range.

ATHYSANUS PUSILLUS (Hook.) Greene. Common vernal annual on grassy slopes across range.

THYSANOCARPUS CURVIPES Hook. Common vernal annual across range on grassy, openly wooded slopes.

THYSANOCARPUS LACINIATUS Nutt. var. CRENATUS (Nutt.) Brew. Habitat and occurrence of preceding.

THYSANOCARPUS RADIANS Benth. Habitat and occurrence of the two species preceding, but much less frequent.

TILLAEA AQUATICA L. var. DRUMMONDII (Torr. and Gray) Jepson. (*T. Drummondii* Torr. and Gray; *Tillaeastrum Drummondii* Britton.) Occasional in moist areas, interior of range.

TILLAEA ERECTA Hook. and Arn. (*T. minima* Miers.) Occasional on grassy or mossy slopes across range.

SEDELLA PENTANDRA H. K. Sharsmith, Madroño **3**:240. 1936. Occasional on shale, slate, or sandstone areas which dry out early; canyon slopes, edges of open chaparral, or margins of small, intermittent streams, interior and east side of range. Type from Arroyo del Puerto, *C. W. and H. K. Sharsmith 1831* (UC!). Since its discovery in the Mount Hamilton Range, this species has been reported in the South Coast Ranges from the Pinnacles of San Benito County (Howell, Leafl. West. Bot. **2**:99. 1938— *Howell 12939*, CA, UC), and from the San Carlos Range (Mason, oral communication, 1940). It has been collected recently in the inner North Coast Ranges in Lake County as well (Sharsmith, Madroño **5**:194. 1940).

SEDUM RADIATUM Wats. Occasional among rocks, Mount Hamilton.

SEDUM SPATHULIFOLIUM Hook. Cliffs, north base Mount Hamilton.

ECHEVERIA LAXA Lindl. var. PANICULATA Jepson; type from Morrison Canyon near Niles (extreme north end of Mount Hamilton Range), *Jepson 13419*. (*Dudleya paniculata* Brit. and Rose; *Cotyledon laxa* var. *paniculata* Jepson.) Occasional on rock cliffs across range. This variety is of dubious distinction. The small stature of the plants and consequent "paniculate" inflorescence may be merely a matter of the growth stage represented. If the plants were to increase in size, they might show a "racemose" type of inflorescence like that attributed to *E. laxa* var. *Setchellii* (see below).

ECHEVERIA LAXA Lindl. var. SETCHELLII Jepson, type from Coyote Creek, *Setchell and Jepson* in 1896 (foothills at west base Mount Hamilton Range). (*Cotyledon laxa* var. *Setchellii* Jepson; *Dudleya Setchellii* Brit. and Rose.) Occasional on cliffs across range. There is apparently no geographic segregation of this variety and the preceding; both as known appear to be limited to the Mount Hamilton Range and the adjacent Santa Clara Valley, and are here regarded as questionably distinct.

SAXIFRAGACEAE

SAXIFRAGA CALIFORNICA Greene. (*S. virginiensis* var. *californica* Jepson; *Micranthes californica* Small.) Occasional on moist, wooded slopes across range.

LITHOPHRAGMA AFFINIS Gray. Occasional to abundant across range on moist, wooded slopes.

LITHOPHRAGMA CYMBALARIA Torr. and Gray. Arroyo del Puerto on moist, rocky slope, *H. K. Sharsmith 1630* (UC). This species has been known hitherto from hills mostly near the coast from Monterey County to San Diego County; the above collection thus represents a long extension of known range.

LITHOPHRAGMA HETEROPHYLLA (Hook. and Arn.) Torr. and Gray var. SCABRELLA (Greene) Jepson. (*L. scabrella* Greene.) Wooded slopes, west side of range.

HEUCHERA MICRANTHA Dougl. var. PACIFICA Rosend. and Butters. Niles Canyon (extreme northwest margin of Mount Hamilton Range), *Mason 3237* (UC). In the central Coast Ranges more typically an element of the outer Coast Ranges.

RIBES AUREUM Pursh var. GRACILLIMUM (Cov. and Britt.) Jepson. Occasional at stream margins, interior of range.

RIBES SANGUINEUM Pursh var. GLUTINOSUM (Benth.) Loud. (*R. glutinosum* Benth.) Wooded slopes, Mount Hamilton.

RIBES MALVACEUM Smith. Occasional on wooded slopes or in chaparral across range.

RIBES QUERCETORUM Greene. Occasional on brushy or wooded canyon slopes, interior and east side of range, its approximate northern limit of distribution.

RIBES CALIFORNICUM Hook. and Arn. Dry, rocky slopes, west side and interior of range.

RIBES AMARUM McCl. Brush slopes, west side of range. Pyramid Rock, *Hendrix 800* (VTM); Smith Creek, *Lundh 37* (VTM); Mount Hamilton, *H. K. Sharsmith 3346* (UC). Mainly found in southern California in the San Bernardino and San Gabriel mountains; occurring northward in the southern Sierra Nevada, and in the Santa Lucia Mountains of the South Coast Ranges. The specimens cited above agree closely not only with *R. amarum*, but also with *R. Menziesii* var. *hystriculum* Jepson (Fl. Calif. **2**:156. 1936) of Mount Diablo. Apparently these two units are closely related. In *R. amarum* the hypanthium is described as longer than broad and the sepals are ligulate; in *R. Menziesii* and allies the hypanthium is described as broader than long, less than half the length of the sepals, and the sepals are lanceolate. The Mount Hamilton Range specimens agree with *R. amarum* in general characteristics of flower and fruit, and have a hypanthium about as broad as long, but at least half the length of the sepals, and the sepals are more or less ligulate.

RIBES SPECIOSUM Pursh. Wooded slopes, west side of range.

PLATANACEAE

PLATANUS RACEMOSA Nutt. Abundant in canyon bottoms in localized areas. Very abundant in Arroyo Mocho at north margin of range; not seen in interior of range or in Arroyo del Puerto.

ROSACEAE

HOLODISCUS DISCOLOR Maxim. Wooded slopes, west side of range.

RUBUS VITIFOLIUS Cham. and Schlecht. Brushy or wooded slopes, west side of range.

POTENTILLA GLANDULOSA Lindl. subsp. TYPICA Keck. (*P. glandulosa* var. *Wrangeliana* Wolf; *Drymocallis glandulosa* (Lindl.) Rydb.) San Antonio Valley.

ROSA CALIFORNICA Cham. and Schlecht. Common in interior of range in canyon bottoms.

ROSA GYMNOCARPA Nutt. Santa Isabella Creek. Predominately an outer Coast Range species in central California; absent or infrequent in the inner South Coast Ranges.

ALCHEMILLA OCCIDENTALIS Nutt. (*A. arvensis* of authors, non Scop.; *A. cuneifolia* Nutt.) Grassy slopes across range, infrequent.

CERCOCARPUS BETULOIDES Nutt. (*C. betulaefolius* Nutt.) Occasional chaparral constituent across range.

ADENOSTOMA FASCICULATUM Hook. and Arn. Dominant chaparral species of interior of range, sometimes forming extensive pure stands.

OSMARONIA CERASIFORMIS (Torr. and Gray) Greene. Occasional element of soft chaparral, west side and interior of range.

PRUNUS EMARGINATA (Dougl.) Walp. (*Cerasus emarginata* Dougl.) On brushy slopes, west side of range.

PRUNUS VIRGINIANA L. var. DEMISSA (Nutt.) Torr. (*Cerasus demissa* Nutt.; *Padua demissa* Roem.) Occasional on north-facing slopes, west side and interior of range.

PRUNUS ILICIFOLIA (Nutt.) Walp. (*Cerasus ilicifolia* Nutt.) Occasional in chaparral as low shrub, or forming trees 20-30 feet high on moist, northfacing slopes; interior of range.

PRUNUS SUBCORDATA Benth. Forming thickets on north-facing slopes, interior of range.

PHOTINIA ARBUTIFOLIA (Ait.) Lindl. An infrequent chaparral component across range.

*PYRUS MALUS L.? Forming a single thicket, west side Mount Hamilton near summit, *H. K. Sharsmith 1090* (UC); sterile.

AMELANCHIER ALNIFOLIA Nutt. Wooded slopes, west side and interior of range.

LEGUMINOSAE

PROSOPIS CHILENSIS (Molina) Stuntz. (*P. juliflora* DC. var. *glandulosa* (Torr.) Cockerell.) Tesla, Corral Hollow, *Mason* in 1935 (UC), *Ferris 7882* (S). A desert species occurring sparingly at north base of Tehachapi Mountains, and at two or three isolated localities in the inner South Coast Ranges, Corral Hollow representing its known northern limit.

THERMOPSIS MACROPHYLA Hook. and Arn. (*T. californica* Wats.; *T. californica* var. *velutina* Greene.) Dry slopes, west side Mount Hamilton.

PICKERINGIA MONTANA Nutt. Near Round Mountain, *Lundh 20* (VTM). To be expected as a frequent xerophytic chaparral element of interior and east side of range, but apparently uncommon.

LUPINUS ALBIFRONS Benth. (*L. albifrons* var. *collinus* Greene; *L. collinus* Heller.) Frequent on dry, exposed slopes across range.

LUPINUS FORMOSUS Greene. (*L. Pendletonii* Heller, Muhl. **2**:295. 1907; type from Mount Hamilton, *Heller 8610*.) Grasslands, San Antonio Valley.

LUPINUS RIVULARIS Dougl. (*L. latifolius* Agardh.) Openly wooded slopes, Mount Hamilton.

LUPINUS BICOLOR Lindl. (*L. bicolor* var. *microphyllus* C. P. Smith.) Frequent vernal annual on grassy slopes across the range.

LUPINUS SUCCULENTUS Dougl. Occasional on hillslopes or in valley flats across range.

LUPINUS DENSIFLORUS Benth. (*L. microcarpus* Sims var. *densiflorus* Jepson.) Frequent on grassy slopes or in flood beds of streams, across range.

LUPINUS MICROCARPUS Sims. (*L. subvexus* C. P. Smith; *L. subvexus* var. *phoeniceus* C. P. Smith, Bull. Torr. Bot. Club **44**:405. 1917, type from west side Mount Hamilton, *Heller 8652.*) Frequent on grassy slopes across range. Following Jepson (Fl. Calif. **2**:278. 1936) *L. subvexus* is here considered as conspecific with the Chilean *L. microcarpus*. Jepson, *op. cit.*, treats *L. densiflorus* as a variety of *L. microcarpus*, but it is here considered as a distinct species (see above). Both *L. densiflorus* and *L. microcarpus* occur abundantly in the Mount Hamilton Range and are easily distinguished in the field and from dried material. *Lupinus densiflorus* is usually larger than *L. microcarpus*, less villous, and with more fistulous stems; the verticels are secund in fruit in *L. densiflorus*, but not in *L. microcarpus*; and the flowers are yellow in *L. densiflorus* (occasionally tinged with pink or blue), and blue in *L. microcarpus*.

*MEDICAGO LUPULINA L. Occasional, west side of range.

*MEDICAGO HISPIDA Gaertn. (*M. denticulata* Willd.) Occasional, west side of range.

*MEDICAGO APICULATA Willd. Arroyo del Puerto.

*MELILOTUS ALBA Desr. Arroyo del Puerto.

TRIFOLIUM FUCATUM Lindl. Occasional along streams, interior of range.

TRIFOLIUM AMPLECTENS Torr. and Gray. Occasional on hillslopes, interior and east side of range.

TRIFOLIUM DEPAUPERATUM Desv. Occasional in valleys or canyon bottoms, interior of range.

TRIFOLIUM TRIDENTATUM Lindl. Frequent on hillslopes and valleys across range. *H. K. Sharsmith 1033* (UC), Copernicus Peak, is remarkable for its very narrow leaflets, and internodes so short that the dry stipules cover the stems; *H. K. Sharsmith 3710* (UC), a later collection from the same locality, is much less extreme as to these points.

TRIFOLIUM OBTUSIFLORUM Hook. (*T. roscidum* Greene.) Occasional in moist areas across range.

TRIFOLIUM WORMSKJOLDII Lehm. (*T. involucratum* Ortega.) Bog, north side Mount Hamilton.

TRIFOLIUM VARIEGATUM Nutt. Moist areas, west side and interior of range.

TRIFOLIUM OLIGANTHUM (Nutt.) Steud. Infrequent across range on wooded slopes.

TRIFOLIUM CYATHIFERUM Lindl. Occasional in flood bed of Santa Isabella Creek. North base Mount Hamilton, *H. K. Sharsmith 3661* (UC); east base Mount Hamilton, *H. K. Sharsmith 1153* (UC); east base Mount Santa Isabella, *H. K. Sharsmith 3733* (UC). The above collections represent the first record of this northern species in the South Coast Ranges.

TRIFOLIUM BARBIGERUM Torr. Grassland, Mount Day Ridge.

TRIFOLIUM MICRODON Hook. and Arn. Dry slopes, interior and west side of range.

TRIFOLIUM MICROCEPHALUM Pursh. Meadowy areas, or wooded slopes, west side of range.

TRIFOLIUM BIFIDUM Gray. Grassy slopes, west side of range.

TRIFOLIUM CILIOLATUM Benth. (*T. ciliatum* Nutt., non Clarke.) Occasional across range on grassy slopes or valley flats.

TRIFOLIUM GRACILENTUM Torr. and Gray. Occasional across range on grassy slopes or valley flats.

TRIFOLIUM MACRAEI Hook. and Arn. Mount Hamilton, *Elmer 4673* (UC). Typically an element of the immediate coast.

TRIFOLIUM DICHOTOMUM Hook. and Arn. Arroyo del Puerto.

TRIFOLIUM DICHOTOMUM var. TURBINATUM Jepson. Grassy slopes or rocky areas, Mount Hamilton. Distinguished from the species by its low stature and turbinate or ovate heads, and according to Jepson (Fl. Calif. **2**:310. 1936) restricted to Mount Hamilton and Mount Tamalpais.

TRIFOLIUM ALBOPURPUREUM Torr. and Gray. Frequent on grassy slopes across range.

TRIFOLIUM OLIVACEUM Greene var. GRISEUM Jepson. Grassland, San Antonio Valley.

LOTUS CRASSIFOLIUS (Benth.) Greene. Grassland, Mount Hamilton.

LOTUS STRIGOSUS (Nutt.) Greene. Rocky, openly wooded slopes, Arroyo del Puerto.

LOTUS MICRANTHUS Benth. Grassy slopes, west side Mount Hamilton.

LOTUS AMERICANUS (Nutt.) Bisch. Occasional on grassy slopes across range.

LOTUS SUBPINNATUS Lag. Common in grassland or in flood beds of streams across range.

LOTUS HUMISTRATUS Greene. Common in grassland or in flood beds of streams across range.

LOTUS RUBRIFLORUS H. K. Sharsmith, Madroño 6:56-58. 1941. Rolling hills at north end Adobe Valley, Mount Hamilton Range, *Carter and Sharsmith 3544* (type UC!), plants in flower; *Carter and Morrison 1403* (UC) topotype, plants in fruit. Known only at the type locality, but within the one known colony the plants are abundant; a distinctive species.

LOTUS SCOPARIUS (Nutt.) Ottley. Infrequent on rocky slopes across range.

PSORALEA MACROSTACHYA DC. (*Hoita macrostachya* Rydb.) Moist areas, near summit Mount Hamilton.

PSORALEA PHYSODES Dougl. (*Hoita physodes* Rydb.) Wooded slopes, west side of range.

PSORALEA CALIFORNICA Wats. (*Pediomelum californicum* Rydb.) Unstable talus, east side of range.

GLYCYRRHIZA LEPIDOTA Pursh. Hillslopes or valley flats, infrequent, interior of range.

ASTRAGALUS OXYPHYSUS Gray. Type from Arroyo del Puerto, Mount Hamilton Range, *Brewer 1259* (isotype UC!). Arroyo del Puerto, *H. K. Sharsmith 1554, 1773* (UC), *Hoover 851* (UC); Crow's Creek, *Elmer 4358* (WSC). The Arroyo del Puerto collections represent topotypes, and are apparently the first recollections from this area since Brewer collected the type in 1862. The Mount Hamilton Range is the known northern limit of the species.

ASTRAGALUS DOUGLASII Gray? Interior of range on grassy slopes. Arroyo Bayo, *H. K. Sharsmith 1901, 3466* (UC). Questionably referrable to *A. Douglasii;* in the absence of mature pods a positive determination is difficult or impossible.

ASTRAGALUS DIDYMOCARPUS Hook. and Arn. Exposed slopes, east side of range.

ASTRAGALUS GAMBELIANUS Sheld. Common across range on grassy slopes.

*VICIA SATIVA L. Grassy slope, Hall's Valley.

VICIA AMERICANA Muhl. var. TRUNCATA (Nutt.) Brewer. (*V. truncata* Nutt.) Occasional on wooded slopes west side and interior of range.

VICIA CALIFORNICA Greene. Wooded slopes, Mount Hamilton. Closely allied to *V. americana* var. *truncata*, but more villous, and the plants usually only 5-11 inches tall instead of 2-3 feet. The large colonies of *V. californica* are uniformly low in the Mount Hamilton area, and easily distinguished from *V. americana* var. *truncata*.

LATHYRUS BOLANDERI Wats. subsp. QUERCETORUM (Heller) Bradshaw. (*L. vestitus* of authors; *L. quercetorum* Heller, Muhl. 2:290, 1907, type from near summit of Mount Hamilton, *Heller 8623.*) Common on wooded slopes, west side of range, extending into the interior in favorable habitats.

LINACEAE

LINUM CLEVELANDII Greene. (*Hesperolinon Clevelandii* Small.) Occasional in large colonies on rocky slopes in chaparral or on unstable talus, serpentine rock, Red Mountains. Arroyo del Puerto, *H. K. Sharsmith 3748, 3788* (UC). *Carter 862* (UC); Adobe Creek, *H. K. Sharsmith 3755* (UC); Colorado Creek, *H. K. Sharsmith 3801* (UC). Hitherto considered as restricted to chaparral serpentine areas of Lake County, where it is most highly developed, and adjacent areas in Napa and Mendocino counties. The Red Mountains collections thus represent a significant extension of known range.

LINUM MICRANTHUM Gray. (*Hesperolinon micranthum* Small.) Occasional in chaparral across range.

LINUM CALIFORNICUM Benth. (*Hesperolinon californicum* Small.) Rocky slopes in shale or serpentine, east side of range.

LINUM SPERGULINUM Gray. Bald Peak, *Dudley 4197* (S).

GERANIACEAE

GERANIUM CAROLINIANUM L. Grassy slopes, west side of the range.

*GERANIUM DISSECTUM L. Grassy areas, west side and interior of range.

ERODIUM MACROPHYLLUM Hook. and Arn. Orestimba Canyon, *Brewer 1280* (UC). According to Jepson (Fl. Calif. 2:407. 1936), this "is an extremely rare plant, both as to stations and individuals."

*ERODIUM BOTRYS (Cav.) Bertol. Grassy hillsides, west side of range.

*ERODIUM MOSCHATUM (L.) L'Her. Across range, but most abundant on the lower, cultivated slopes of west side.

*ERODIUM CICUTARIUM (L.) L'Her. Common in depauperate form on dry and sparsely grassy hillsides across range.

LIMNANTHACEAE

LIMNANTHES DOUGLASII R. Br. (*Floerkea Douglasii* Baill.) Occasional in moist valley flats, west side and interior of range.

EUPHORBIACEAE

EREOMOCARPUS SETIGERUS (Hook.) Benth. (*Croton setigerus* Hook.) Common autumnal species of the dryer plains and hillsides across range.

EUPHORBIA SERPYLLIFOLIA Pers. (*Chamaesyce serpyllifolia* Small.) Occasional, valleys and hillsides, interior and east side of range.

EUPHORBIA OCELLATA Dur. and Hilg. var. TYPICA Wheeler. Canyon slopes or dry creek beds, east side of range.

EUPHORBIA DICTYOSPERMA Fisch. and Mey. Openly wooded slopes, occasional across range.

EUPHORBIA CRENULATA Engelm. Moist sand, Santa Isabella Creek.

CALLITRICHACEAE

CALLITRICHE MARGINATA Torr. Sag pond, Arroyo Mocho.

CALLITRICHE PALUSTRIS L. Sag pond, Mount Day Ridge.

ANACARDIACEAE

TOXICODENDRON DIVERSILOBUM (Torr. and Gray) Greene. (*Rhus diversiloba* Torr. and Gray). Frequent across the range on canyon slopes.

ACERACEAE

ACER MACROPHYLLUM Pursh. In ravines or at stream margins, west side of range.

SAPINDACEAE

AESCULUS CALIFORNICA (Spach) Nutt. Occasional on north facing slopes across range.

RHAMNACEAE

RHAMNUS CALIFORNICA Esch. subsp. TOMENTELLA (Benth.) Wolf. (*R. tomentella* Benth.; *R. californica* var. *tomentella* Brew. and Wats.) Common on open slopes or sometimes in chaparral, west side and interior of range. Distinguished from the subspecies *typica* Wolf by narrowly elliptical leaves which are tomentose beneath; characteristic of the inner Coast Ranges whereas subspecies *typica* is more maritime. On Mount Hamilton, however, a form occurs which is intermediate between subspecies *typica* and subspecies *tomentella*. It is found only in dense shade, and has the broad, thin leaves of subspecies *typica*; the leaves appear almost glabrous beneath, but when carefully examined they reveal a slight tomentum such as is characteristic of subspecies *tomentella*. The following collection represents this shade form: Santa Isabella Creek, *H. K. Sharsmith 3659* (UC), undershrub beneath dense growth of *Acer macrophyllum* and *Arbutus Menziesii*. A similar shade form was noted by Greene (Erythea 1: 82. 1893) at Joaquin (Murietta) Springs.

RHAMNUS CROCEA Nutt. subsp. ILICIFOLIA (Kell.) Wolf. (*R. ilicifolia* Kell.; *R. crocea* var. *ilicifolia* Greene.) On wooded slopes or in chaparral, occasional across range.

CEANOTHUS INTEGERRIMUS Hook. and Arn. Single shrub in dense shade, Santa Isabella Creek, *H. K. Sharsmith 3660* (UC).

CEANOTHUS LEUCODERMIS Greene. (*C. divaricatus* of authors, non Nutt.) Chaparral constituent of interior and east side of range, most common in Red Mountains area.

CEANOTHUS SOREDIATUS Hook. and Arn. Infrequent chaparral constituent across range. A specimen from Mount Sizer, *Hendrix 747* (VTM), has leaf characteristics of both *C. sorediatus* and *C. dentatus* Torr. and Gray; according to the collector's note it appeared to be a hybrid between these two species, but the material is without flowers or fruits and positive determination is difficult.

CEANOTHUS CUNEATUS (Hook.) Nutt. Next to *Adenostoma fasciculatum* the most abundant chaparral element; interior and east side of range. Greene (Erythea 1:79. 1893) comments that the genus *Ceanothus* "seems to be wholly absent from Mount Hamilton." Greene must have confined his botanical activities on Mount Hamilton entirely to the west side and summit areas of the mountain, for *C. cuneatus* is extremely abundant on the east slope below 3500 feet.

CEANOTHUS FERRISAE McMinn, Madroño 2:89. 1933. Type from above Coyote Creek, Madrone Springs road, Mount Hamilton Range, *Abrams 6626* (S!). Known only from the west slope of the Mount Hamilton Range near Madrone Springs, and from a few stations in the Santa Cruz Mountains; it is related to *C. cuneatus*, but is distinguished by larger, more elliptical, and toothed leaves.

VITACEAE

VITIS CALIFORNICA Benth. Isolated localities, west side of range.

MALVACEAE

SIDALCEA DIPLOSCYPHA (Torr. and Gray) Gray. Wooded areas, east side Mount Hamilton.

SIDALCEA MALVAEFLORA (DC.) Gray. Grassland near Hall's Valley.

MALVASTRUM FREMONTII Torr. var. CERCOPHORUM Rob., type from Arroyo del Vallé, Mount Hamilton Range, *Greene*. Occasional on dry slopes of interior, more abundant on east side of range.

MALVASTRUM PARRYI Greene. Corral Hollow.

ELATINACEAE

ELATINE BRACHYSPERMA Gray. Sag pond, Mount Day Ridge.

ELATINE CALIFORNICA Gray. Sag pond, Mount Day Ridge.

VIOLACEAE

VIOLA DOUGLASII (Hook.) Steud. Frequent in early spring on grassy slopes, west side and interior of range.

VIOLA SHELTONII Torr. Rocky chaparral slope, near summit of Copernicus Peak, *H. K. Sharsmith 1040* (UC). A fragmentary specimen collected after the plants had shed their seeds, but leaf shape and underground parts make its identity unquestionable. *Viola Sheltonii* has not been reported before from the South Coast Ranges except from Mount Diablo. It occurs at medium elevations in the North Coast Ranges, Sierra Nevada, and in the mountains of southern California, and might well be expected in the South Coast Ranges.

VIOLA PURPUREA Kell. Frequent on rocky, openly wooded slopes across range.

VIOLA PEDUNCULATA Torr. and Gray. Open grassy slopes, interior and west side of range.

LOASACEAE

MENTZELIA LAEVICAULIS (Dougl.) Torr. and Gray. Dry floodbeds, occasional across range.

MENTZELIA LINDLEYI Torr. and Gray subsp. TYPICA Wolf. (*M. Lindleyi* Torr. and Gray; *Acrolasia aurea* Rydb.) Dry, rocky slopes across range, but most abundant in interior and on east side. The subspecies reaches its highest development in the Mount Hamilton Range, presumably the type locality according to Wolf (Occasional Papers, Rancho Santa Ana Bot. Gard. series 1, No. 2. 69-73. 1938).

MENTZELIA MICRANTHA (Hook. and Arn.) Torr. and Gray. Occasional on open slopes, west side of range.

MENTZELIA DISPERSA Wats. Dry open slopes or in chaparral, interior and east side of range.

MENTZELIA GRACILIENTA Torr. and Gray. Canyon slopes, east side of range. A southern California and Great Basin species which reaches its northern limit in the Mount Hamilton Range.

DATISCACEAE

DATISCA GLOMERATA (Presl.) Baill. Occasional; wooded slopes, Mount Hamilton.

LYTHRACEAE

LYTHRUM ADSURGENS Greene.. Moist sand, Santa Isabella Creek.

ONAGRACEAE

ZAUSCHNERIA CALIFORNICA Presl. Rocky slopes, occasional across range.

EPILOBIUM MINUTUM Lindl. Occasional, interior and east side of range, in flood beds of streams or on talus.

EPILOBIUM PANICULATUM Nutt. Occasional on grassy slopes across range.

BOISDUVALIA DENSIFLORA (Lindl.) Wats. (*B. bipartita* Greene, Erythea 3:119. 1895; type from Arroyo del Vallé, Mount Hamilton Range, *Greene.*) Occasional in moist areas, west side and interior of range.

BOISDUVALIA STRICTA (Gray) Greene. Moist gully, San Antonio Valley.

CLARKIA RHOMBOIDEA Dougl. Wooded slopes, east side Mount Hamilton.

CLARKIA ELEGANS Dougl. Frequent across range on exposed slopes.

CLARKIA CONCINNA (Fisch. and Mey.) Greene. Moist wooded slopes, west side and interior of range.

CLARKIA BREWERI (Gray) Greene. (*C. Saxeana* Greene.) Rocky slopes mainly of unstable shale talus; summit areas across range, most abundant on east side. Type from Mount Oso, Mount Hamilton Range, *Brewer.* In the Mount Hamilton Range the colonies of this central Coast Range endemic are highly localized and usuualy small, although the individuals may be very abundant within a colony. When the plants are in full bloom some of the colonies high on the steep shale slopes of Arroyo del Puerto can be seen at a distance of one-half mile or more, due to the abundance of plants within the colony and the vivid pink of the large flowers.

GODETIA AMOENA (Lehm.) Don. (*G. rubicunda* Lindl.) Frequent in grassland, west side and interior of range.

GODETIA QUADRIVULNERA (Dougl.) Spach. Dry hillsides, west side and interior of range.

GODETIA QUADRIVULNERA var. ELMERI Jepson. (*G. purpurea* Don var. *Elmeri* Jepson; *G. purpurea* var. *parviflora* (Wats.) Hitchc.) Occasional, grassy slopes, west side of range.

GODETIA EPILOBIOIDES (Nutt.) Wats. (*Clarkia epilobioides* Nels. and Macbr.; *G. epilobioides* var. *modesta* (Jepson) Jepson.) Wooded slopes, interior and east side of range.

OENOTHERA DELTOIDES Torr. and Frem. var. COGNATA (Jeps.) Munz, type from Corral Hollow, Mount Hamilton Range, *Brewer 1217.* Corral Hollow.

OENOTHERA MICRANTHA Hornem. var. JONESII (Levl.) Munz. (*Sphaerostigma micranthum* var. *Jonesii* Nelson; *Oenothera hirtella* Greene.) Occasional in chaparral or rocky flood beds, interior and east side of range. Variety *typica* Munz is found mainly on the immediate coast, but Munz (Bot. Gaz. **85**:262. 1928), in discussing the frequent intergradations, cites the following as an intergrade between variety *typica* and variety *Jonesii*: Mount Hamilton-Livermore road, *Bacigalupi* in 1923 (S).

OENOTHERA CONTORTA Dougl. var. STRIGULOSA (Fisch. and Mey.) Munz. Dry slopes or in chaparral, interior and east side of range.

OENOTHERA DECORTICANS (Hook. and Arn.) Greene var. TYPICA Munz. *O. decorticans* of authors, *O. alyssoides* var. *decorticans* Jepson.) Unstable talus, east side of range. A desert dwelling species which reaches its northern extremity in the Mount Hamilton Range.

ARALIACEAE

ARALIA CALIFORNICA Wats. Wooded gullies below springs, east slope Mount Hamilton. Characteristic of the outer Coast Ranges rather than the inner.

UMBELLIFERAE

SANICULA CRASSICAULIS Poepp. (*S. Menziesii* Hook. and Arn.) West slope Mount Hamilton.

SANICULA BIPINNATIFIDA Dougl. Occasional on open grassy slopes across range.

SANICULA BIPINNATA Hook. and Arn. Occasional on open grassy slopes, west side and interior of range.

SANICULA TUBEROSA Torr. Occasional on openly wooded slopes, west side and interior of range.

SANICULA SAXATILIS Greene. Rocky opening in chaparral, northeast ridge of Copernicus Peak, *H. K. Sharsmith 923* (UC); unstable talus, east side Mount Hamilton, *Bowerman 924* (UC), *H. K. Sharsmith 1925* (UC). *Sanicula saxatilis* Greene is among the least known and most narrowly restricted of the species endemic to the central Coast Ranges. For many years this distinctive species was known only from rocky crests near the summit of Mount Diablo, the northernmost peak of the South Coast Ranges, where it was first collected by Greene in 1893. Only three or four collections have been made from Mount Diablo since. *Sanicula saxatilis* was found first on Mount Hamilton by Bowerman. In the two Mount Hamilton localities known for the species, the plants of the colony grow within a sharply limited area; the individuals are scattered and not abundant, and are restricted to almost barren, talus "islands" in the chaparral. The fleshy roots are tuberous, and are wedged between the rocks and often much distorted.

OSMORHIZA BRACHYPODA Torr. Infrequent on wooded slopes, Mount Hamilton.

OSMORHIZA NUDA Torr. Wooded slopes, west side of range.

DAUCUS PUSILLUS Michx. Infrequent on rocky, openly wooded slopes across range.

APIASTRUM ANGUSTIFOLIUM Nutt. Infrequent, east side of range.

CAUCALIS MICROCARPA Hook. and Arn. Occasional across range on grassy slopes.

BOWLESIA INCANA Ruiz and Pavon. (*B. septentrionalis* Coult. and Rose.) Infrequent on moist shaded banks of streams across range.

*CONIUM MACULATUM L. In dense stands along roadside where natural vegetation was cleared by burning, Grand View.

PERIDERIDIA GAIRDNERI (Hook. and Arn.) Mathias. (*Carum Gairdneri* Gray.) Steep, rocky slope, Santa Isabella Valley.

PERIDERIDIA CALIFORNICA (Torr.) Nels. and Macbr. (*Eulophus californicus* Coult. and Rose.) Frequent along stream margins, interior of range.

DEWEYA HARTWEGI Gray. (*Velaea Hartwegii* Coult. and Rose.) Infrequent; wooded slopes, west side and interior of range. A widely distributed, but locally rare species.

DEWEYA KELLOGGII Gray. (*Velaea Kelloggii* Coult. and Rose.) Wooded, rocky slopes, Arroyo Bayo. Like the preceding, widely distributed but locally rare, the known stations scattered.

LEPTOTAENIA CALIFORNICA Nutt. Occasional on wooded or brushy slopes, interior of range.

LOMATIUM CARULIFOLIUM (Hook. and Arn.) Coult. and Rose. (*Peucedanum carulifolium* Torr. and Gray.) Rocky summit areas, Mount Hamilton.

LOMATIUM UTRICULATUM (Nutt.) Coult. and Rose. Occasional on rocky, openly wooded slopes across the range.

LOMATIUM MACROCARPUM (Hook. and Arn.) Coult. and Rose. Common on openly wooded, grassy slopes, interior and east side of range.

LOMATIUM DASYCARPUM (Torr. and Gray) Coult. and Rose. (*Peucedanum tomentosum* acc. Greene, Erythea 1:88. 1893, non Benth.) Occasional in dry open slopes across range.

LOMATIUM NUDICAULE (Pursh) Coult. and Rose. Occasional in interior of range on dry, open slopes.

ANGELICA TOMENTOSA Wats. Moist soil, spring margins, east slope of Mount Hamilton.

ERYNGIUM VASEYI Coult. and Rose var. CASTRENSE (Jepson) Hoover ex Mathias and Constance. Occasional in drying pools, interior of range.

GARRYACEAE

GARRYA FREMONTII Torr. Chaparral slopes, summits of peaks on west side of range and through interior. The Mount Hamilton Range material is atypical according to Bacigalupi in that the leaves are hairy beneath, with straight, appressed hairs, rather than glabrous.

GARRYA CONGDONI Eastw. (*G. flavescens* var. *venosa* Jepson.) Canyon slopes of serpentine rock, west side of range and in Red Mountains. A little known species restricted to the inner Coast Ranges from Tehama County and Lake County to San Benito County, also in the foothills of the central Sierra Nevada. Only on serpentine in the Mount Hamilton Range.

CORNACEAE

CORNUS GLABRATA Benth. In thickets along stream margins, west side and interior of range.

CORNUS STOLONIFERA Michx. var. CALIFORNICA (C. A. Mey.) McMinn. (*C. pubescens* var. *californica* Coult. and Evans; *C. californica* var. *pubescens* Macbr.) Occasional at stream margins, west side and interior of range.

ERICACEAE

ARBUTUS MENZIESII Pursh. Occasional, wooded slopes on west side of range. An outer Coast Range species which is rare in the inner Coast Ranges.

ARCTOSTAPHYLOS GLAUCA Lindl. (*A. manzanita* Parry acc. Greene, Erythea 1:92. 1893.) Frequent in chaparral across range.

ARCTOSTAPHYLOS GLANDULOSA Eastw. var. CAMPBELLAE (Eastw.) Adams. (*A. Campbellae* Eastw., type from Mount Hamilton, *Campbell* in 1922 (CA); *A. tomentosa* Dougl. acc. Greene, Erythea 1:92. 1893.) Frequent on summit areas of west side of range, occasional in chaparral of interior; restricted to the Mount Hamilton Range.

PRIMULACEAE

DODECATHEON HENDERSONII Gray. Frequent on openly wooded slopes across range.

DODECATHEON HENDERSONII Gray var. BERNALINUM (Greene) Jepson. Hillsides and valley flats across range, most frequent in interior; flowering in early spring, the dense colonies often covering large areas. Mount Hamilton, *H. K. Sharsmith 530* (UC); Arroyo Mocho, *H. K. Sharsmith 1460* (UC); Arroyo del Puerto, *H. K. Sharsmith 1654* (UC). According to H. L. Mason (oral communication), this variety is more properly placed under *D. Clevelandii* Greene, which it resembles except for smaller habit, and shorter, more obtuse anthers.

ANDROSACE OCCIDENTALIS Pursh var. ACUTA (Greene) Jepson. (*A. acuta* Greene.) Rocky areas, interior and east side of range; infrequent. Arroyo Bayo, *H. K. Sharsmith 3067* (UC); Red Mountains, *H. K. Sharsmith 1681* (UC); Arroyo del Puerto, *H. K. Sharsmith 1585* (UC). Widely distributed but rare, the few known stations very isolated; not before recorded from the Mount Hamilton Range. This variety is recognized as a distinct species by Munz (Man. S. Calif. Bot. 371. 1935), but in such a polymorphic genus as *Androsace*, the distinctions between *A. occidentalis* of the eastern states and this Californian representative do not seem sufficient to warrant specific segregation. In variety *acuta* the plants are more delicate, the umbels have fewer flowers, the pedicels are longer, and the calyx teeth are much narrower, almost subulate, but *A. occidentalis* grades toward the variety in all these characteristics.

*ANAGALLIS ARVENSIS L. Grasslands, west slope Mount Hamilton.

OLEACEAE

FRAXINUS DIPETALA Hook. and Arn. Headwaters Arroyo del Puerto.

FORESTIERA NEOMEXICANA Gray. Forming thickets near stream margin, headwaters Arroyo Mocho, *H. K. Sharsmith 1715* (UC). Mainly of the southwestern United States, occurring sparingly northward in the inner South Coast Ranges to the Mount Hamilton Range.

GENTIANACEAE

CENTAURIUM FLORIBUNDUM Rob. (*Erythraea floribundum* Benth.) Grassy slopes, east side of range.

APOCYNACEAE

APOCYNUM CANNABINUM L. var. GLABERRIMUM DC. Occasional at stream margins, interior of range.

ASCLEPIADACEAE

ASCLEPIAS MEXICANA Cav. Occasional on dry hills, west side and interior of range.

ASCLEPIAS CALIFORNICA Greene. Sparingly distributed across the range cn open, dry chaparral slopes.

CONVOLVULACEAE

CONVOLVULUS MALACOPHYLLUS Greene. (*C. villosus* (Kell.) Gray, non Pers.; *C. villosus* var. *pedicellata* Jepson; *C. collinus* of authors, non Greene.) Infrequent across range on rocky slopes in chaparral, areas of serpentine rock. The South Coast Range material of *C. malacophyllus* corresponds to *C. villosus* var. *pedicellata* Jepson. Although there is geographical segregation, the variety has a minor morphological basis. If it were to be recognized, a new combination under *C. malacophyllus* would be necessary.

CONVOLVULUS CALIFORNICUS Choisy. (*C. subacaulis* (Hook. and Arn.) Greene, non Buch.-Ham ex Wall.) Dry slopes, west side of range.

CUSCUTA SUBINCLUSA Dur. and Hilg. Parasitic on various shrubs, occasional across range.

POLEMONIACEAE

COLLOMIA GRANDIFLORA Dougl. Occasional on wooded slopes, west side of range.

COLLOMIA DIVERSIFOLIA Greene. Serpentine talus Adobe Creek canyon, *H. K. Sharsmith 3584* (UC). Known hitherto by only four collections from Mendocino, Colusa, and Lake Counties of the North Coast Ranges. This represents one of the species common to serpentine talus in Lake County and in the Red Mountains of the Mount Hamilton Range, but not occurring between.

PHLOX GRACILIS (Dougl.) Greene. (*Gilia gracilis* Hook.) Frequent spring annual, grassy slopes across range.

NAVARRETIA ABRAMSI Elmer. Frequent in rocky soil of chaparral areas, interior and east side of range. The Mount Hamilton Range collections add greatly to our knowledge of this previously little known species, as discussed in a separate paper (Sharsmith, H. K., Amer. Midl. Nat. **32**:510-512. 1944.)

NAVARRETIA PUBESCENS (Benth.) Hook. and Arn. Occasional in grasslands across range.

NAVARRETIA INTERTEXTA (Benth.) Hook. (*Gilia intertexta* Steud.) Occasional near stream margins or at vernal pools, interior of range.

NAVARRETIA MELLITA Greene. Occasional in chaparral, interior of range.

HUGELIA FILIFOLIA (Nutt.) Jepson. (*Gilia filifolia* Nutt. var. *typica* Craig.) Chaparral ridge between Arroyo Bayo and San Antonio Valley, *H. K. Sharsmith 3299* (UC). A widely ranging species, found throughout the more arid portions of western United States. The Mount Hamilton Range material corresponds to the typical phase of the species, that common to coastal southern California from Santa Barbara County to northern Lower California. The above collection and a previously unrecorded one from Lake County of the North Coast Ranges (Near Kelseyville, *Schulthess* in 1931, UC) represent long extensions northward of the species' known range.

Hugelia filifolia is closely related to *Navarretia Abramsii*. In the Mount Hamilton Range *H. filifolia* was found in the same type of habitat as that which *N. Abramsii* frequents, and *N. Abramsii* was collected from the same locality as the collection *3299* cited above. The plants of *H. filifolia*, although superficially similar to those of *N. Abramsii*, were distinguished in the field by their taller size, less branched and more erect stems, and narrower flower clusters.

HUGELIA PLURIFLORA (Heller) Ewan. (*Gilia virgata* var. *floribunda* Gray; *Gilia pluriflora* Heller; *Hugelia Brauntonii* (Jepson and Mason) Jepson.) Steep talus or rocky chaparral areas, east side of the range.

GILIA ACHILLEAEFOLIA Benth. Openly wooded slopes, occasional across range.

GILIA TRICOLOR Benth. Frequent vernal annual on open hillsides or valley flats across range.

GILIA SP. Grassy slopes, west side and interior of range. Smith Creek, *H. K. Sharsmith 614, 1017* (UC); Packard Ridge, *Mason 7203, 7204* (UC). This is a species of infrequent occurrence in the Coast Ranges, for which only a manuscript name is available. It is related to G. *tricolor* and G. *multicaulis*, but is probably closer to the latter.

GILIA MULTICAULIS Benth. (G. *peduncularis* Eastw.) Frequent on open hillsides across range. Although *Gilia peduncularis* is often given specific rank, it is here treated as synonymous with G. *multicaulis*, for the Mount Hamilton Range material shows every intergradation between the cymosely clustered, loose glomerules of flowers in the inflorescence of typical G. *multicaulis*, and the pedicelled, solitary flowers in the open, cymose inflorescence of typical G. *peduncularis*.

GILIA MILLEFOLIATA Fisch. and Mey. Infrequent across range on open hillsides. Often intergrading with G. *multicaulis*, but appearing to be much more distinct than G. *peduncularis*.

GILIA GILIOIDES (Benth.) Greene. Frequent vernal annual on dry slopes across range.

LINANTHUS DICHOTOMUS (Benth.) Benth. (*Gilia dichotoma* Benth.) Occasional on open hillsides across range.

LINANTHUS PHARNACEOIDES (Benth.) Greene. (*L. liniflorus* (Benth.) Greene.) Occasional, west side of range.

LINANTHUS AMBIGUUS (Rattan) Greene. (*Gilia ambigua* Rattan; *Dactylophyllum ambiguum* Heller; type from Oak Hill, west base Mount Hamilton Range, Santa Clara County, *Rattan*.) Frequent on open hillsides or gravelly flats across range.

LINANTHUS PYGMAEUS (Brand) Howell. (*Gilia pusilla* and *Linanthus pusilla* of authors, non Greene.) Open slopes or gravelly flats, interior and east side of range.

LINANTHUS DENSIFLORUS (Benth.) Milliken. (*L. grandiflorus* Greene.) Open slopes, west side and interior of range near its western margin.

LINANTHUS ANDROSACEUS (Benth.) Greene. Common on open slopes across range.

LINANTHUS PARVIFLORUS (Benth.) Greene. (*Gilia lutea* Steud.) Frequent on open slopes, interior and east side of range.

LINANTHUS BICOLOR (Nutt.) Greene. (*Gilia bicolor* Brand.) Frequent on open slopes across range. This species and the preceding form a complex, the relationships of which are not clear.

LINANTHUS CILIATUS (Benth.) Greene. (*Gilia ciliata* Benth.) Occasional in grassland, interior and east side of range.

HYDROPHYLLACEAE

HYDROPHYLLUM OCCIDENTALE Gray. Occasional on brushy slopes, interior of range.

NEMOPHILA MENZIESII Hook. and Arn. (*N. insignis* Benth.) Occasional early spring annual on wooded slopes across range.

NEMOPHILA PEDUNCULATA Dougl. Occasional on wooded slopes, west side and interior of range.

NEMOPHILA SEPULTA Parish. (*N. pedunculata* var. *sepulta* Nels. and Macbr.) Occasional on openly wooded slopes, west side and interior of range. Questionably distinct from the preceding species.

NEMOPHILA HETEROPHYLLA Fisch. and Mey. var. NEMORENSIS (Eastw.) Jepson. Occasional, west side of range on grassy slopes. The variety doubtfully distinguishable from the species.

PHOLISTOMA MEMBRANACEA (Benth.) Constance. (*Ellisia membranacea* Benth., and var. *hastifolia* Brand.) Occasional in moist areas, east side and interior of range.

PHOLISTOMA AURITA (Lindl.) Lilja. (*Nemophila aurita* Lindl.) Shaded canyons, west and north margins of the range.

EUCRYPTA CHRYSANTHEMIFOLIA (Benth.) Greene. (*Ellisia chrysanthemifolia* Benth.) Shaded areas, west side and north margin of range.

PHACELIA CALIFORNICA Cham. var. IMBRICATA (Greene) Jepson. (*P. imbricata* Greene; *P. californica* Cham. var. *calycosa* (Gray) Dundas.) Occasional across range on rocky, dry slopes.

PHACELIA STIMULANS Eastw. Restricted to unstable talus or rocky slopes, Red Mountains. East edge San Antonio Valley, *C. W. and H. K. Sharsmith 982* (UC); Colorado Creek, *H. K. Sharsmith 3894* (UC); Arroyo del Puerto, *H. K. Sharsmith 3786* (UC). A segregate of the *P. magellanica* complex, and like the latter species a complex entity. Specimens determined by J. T. Howell. In *P. magellanica* the corolla is tubular-campanulate to open-campanulate with the lobes suberect or distinctly spreading. In *P. stimulans* the corolla is tubular with lobes erect only at anthesis and curving inward thereafter so that the tips are approximate; in addition the sepals are widest at or above the middle, and the glandular hairs are usually numerous on stems, leaves, and sepals.

PHACELIA TANACETIFOLIA (Benth.) (*P. tanacetifolia* var. *pseudo-distans* Band, type from Red Mountains, Mount Hamilton Range, *Elmer 4338*). Across range on rocky slopes, but infrequent.

PHACELIA BREWERI Gray. Frequent on rocky ridges or loose shale slopes across range. This central Coast Range endemic species is very abundant in the Mount Hamilton Range in suitable habitats at elevations from 250 to 4000 feet.

PHACELIA DISTANS Benth. (*P. leptostachya* Greene; *P. distans* var. *australis* Brand.) Frequent across range on exposed hillsides.

PHACELIA RAMOSISSIMA Dougl. var. SUFFRUTESCENS Parry. Occasional on openly wooded slopes, west side and interior of the range.

PHACELIA CILIATA Benth. In dried-out vernal pools, east side of range.

PHACELIA RATTANII Gray. Occasional on shaded hillslopes or near streams, interior of range. Santa Isabella Creek, *C. W. and H. K. Sharsmith 3732* (UC); San Antonio Creek, *C. W. and H. K. Sharsmith 996* (UC). The Santa Cruz Mountains and Mount Hamilton Range (the above specimens mark the first record from this latter area) represent the known southern distribution of this North Coast Range and southern Oregon species.

PHACELIA DOUGLASII (Benth.) Torr. var. PETROPHILA Jepson, type from Corral Hollow, Mount Hamilton Range, *Jepson 9583*. Corral Hollow.

PHACELIA DIVARICATA (Benth.) Gray. Occasional on open rocky or grassy slopes, interior of range.

PHACELIA PHACELIOIDES (Benth.) Brand. (*P. circinatiformis* Gray.) Infrequent on rocky slopes, interior of range. Mount Hamilton, *H. K. Sharsmith 3724* (UC); Sugarloaf Mountain, *H. K. Sharsmith 3640* (UC); Sweetwater Creek, *C. W. and H. K. Sharsmith 3080* (UC). For many years this species was known only from the original Douglas collections ("California"—exact locality uncertain). T. Brandegee cited it (Zöe **4**: 115. 1893) from Mount Hamilton (*W. W. Price* in 1890) as *P. circinatiformis*, but the Price specimen cannot now be located. In more recent years several collections were made from Mount Diablo. The collections, therefore, amplify considerably the available material of this species. It occurs as low as 2100 feet both on Mount Diablo and in the Mount Hamilton Range.

PHACELIA FREMONTII Torr. Corral Hollow, *Hoover 3032* (UC). A typically desert species which reaches its northern limit in the Mount Hamilton Range.

LEMMONIA CALIFORNICA Gray. In chaparral. Arroyo Mocho, *H. K. Sharsmith 3511* (CA); west slope Red Mountains, *H. K. Sharsmith 3619* (UC). The previously recorded distribution of *L. californica* is as follows: Mohave Desert, to Tehachapi Mountains and southern Sierra Nevada, south to northern Lower California; Lake County of North Coast Ranges. The above collections are thus the first to be recorded from the South Coast Ranges, and they render much less discontinuous the known distribution of this essentially desert species.

EMMENANTHE PENDULIFLORA Benth. Occasional on hillsides or in sandy flood beds of streams, interior and east side of range.

EMMENANTHE PENDULIFLORA var. ROSEA Brand. Steep serpentine talus, Red Mountains. Arroyo del Puerto, *Mason* in 1935 (UC), *C. W. and H. K. Sharsmith 3145* (UC); Adobe Creek, *H. K. Sharsmith 3781* (UC); Red Mountain, *Elmer 4877* (UC). The type locality of this little known variety is Mount Pinos, Tehachapi Mountains, Ventura County, California, and its known distribution is limited to Mount Pinos and to the South Coast Ranges as far north as the Red Mountains of the Mount

Hamilton Range. The only character used by Brand (Pflanzr. 4(251):134. 1913) to distinguish the variety *rosea* is the pink flowers as compared to the yellow flowers of the species, and subsequent workers have usually reduced the variety to synonymy under the species. In the Red Mountains, however, the variety *rosea*, in addition to the pink flowers, shows the following points of differentiation from the species: leaves fern-like, more finely and regularly dissected than in the species, sometimes pinnate or with but a narrow strip of leaf blade on either side of the midrib between the widely spaced lobes; stems reddish (green in the species); paniculate cymes longer, flowers more widely spaced on the rachis than in the species. There seems to be some intergradation with the species in all these characters, but on the basis of the above the variety *rosea* appears to be well marked and fully worthy of recognition.

There is a lack of geographical segregation between species and variety, but in the Mount Hamilton Range there is indication of an ecological separation. Although the species occurs from one margin of the range to the other in its typical phase, it is absent from the Red Mountains where the variety is found. The Red Mountains consist of decomposed serpentine rock, and it is possible that the variety *rosea* may be limited throughout its range to areas of serpentine rock.

ERIODICTYON CALIFORNICUM (Hook. and Arn.) Greene. Occasional chaparral constituent, interior and east side of range.

BORAGINACEAE

HELIOTROPIUM CURASSAVICUM L. var. OCULATUM (Heller) Johnston. (*H. curassivicum* L. in part.) Flood beds of streams, interior and east side of range; infrequent.

CYNOGLOSSUM GRANDE Dougl. Wooded slopes, west side and interior of range.

PECTOCARYA LINEARIS DC. var. FEROCULA Johnston. Corral Hollow.

PECTOCARYA PUSILLA (DC.) Gray. Wooded areas, interior of range.

PECTOCARYA PENICILLATA (Hook. and Arn.) DC. Hillslopes, east side of range.

PECTOCARYA SETOSA Gray. Corral Hollow, *Hoover 3042* (UC). A typically desert and Great Basin species, not before found as far north as the Mount Hamilton Range in the inner South Coast Ranges.

AMSINCKIA GRANDIFLORA Kleeb. (*A. spectabilis* of some authors, non Fisch. and Mey.) Corral Hollow, *Hoover 2866, 3021, 3357* (UC). A seldom collected, distinctive and highly restricted species, known only from Antioch, Contra Costa County, and the northern end of the Mount Hamilton Range in the vicinity of Corral Hollow.

AMSINCKIA VERNICOSA Hook. and Arn. Corral Hollow, Hospital Canyon. A typically desert species of infrequent occurrence in the South Coast Ranges to Corral Hollow.

AMSINCKIA TESSELATA Gray. Occasional, east side of range.

AMSINCKIA PARVIFLORA Heller (Muhl. 2:313. 1907), type from Alum Rock Park, *Heller 8470*. Occasional, west side and interior of range.

AMSINKIA INTERMEDIA Fisch. and Mey. (*A. spectabilis* and *A. Douglasiana* of authors.) Occasional across range on grassy slopes.

AMSINCKIA EASTWOODAE Macbride. (*A. Douglasiana* var. *Eastwoodae* Jepson.) Arroyo del Puerto.

CRYPTANTHE FLACCIDA (Dougl.) Greene. Frequent on dry, open hillsides across range.

CRYPTANTHE HISPIDISSIMA Greene. Arroyo del Puerto, *C. W. and H. K. Sharsmith 1537a* (UC), immature.

CRYPTANTHE NEVADENSIS Nels. and Kenn. var. RIGIDA Johnston. Open hillsides or grassy flats, east margin of range.

CRYPTANTHE TORREYANA Greene var. PUMILA (Heller) Johnston. (*C. pumila* Heller.) Open hills, chaparral slopes, or valley flats across range.

CRYPTANTHE SPARSIFLORA Greene. Arroyo del Puerto.

CRYPTANTHE NEMACLADA Greene. Arroyo del Puerto.

CRYPTANTHE CLEVELANDII Greene. Hospital Canyon.

CRYPTANTHE COROLLATA (Johnston) Johnston. (*C. decipiens* var. *corollata* Johnston.) Arroyo del Puerto.

PLAGIOBOTHRYS BRACTEATUS (Howell) Johnston. (*Allocarya bracteata* Howell;

A. californica of authors.) Frequent at stream margins or edges of vernal pools, west side and interior of range.

PLAGIOBOTHRYS ACANTHOCARPUS (Piper) Johnston. (*Allocarya acanthocarpa* Piper; *Plagiobothrys Greenei* (Gray) Johnston; *Echinoglocin acanthocarpa* Brand.) Open grassland, east margin of range.

PLAGIOBOTHRYS TENELLUS (Nutt.) Gray. Frequent on open hillsides across range.

PLAGIOBOTHRYS MYOSOTOIDES (Lehm.) Brand. Chaparral ridge between Santa Isabella Valley and Arroyo Bayo, *C. W. and H. K. Sharsmith 1893* (UC). A South American species, of which this is the first North American collection; cited by I. M. Johnston (Journ. Arn. Arb. **20**:381. 1939).

PLAGIOBOTHRYS CANESCENS Benth. Grassy slopes, east side of range and in Arroyo Mocho.

PLAGIOBOTHRYS ARIZONICUS (Gray) Greene. Corral Hollow. A typically desert borage for which only isolated stations are known in the inner South Coast Ranges.

PLAGIOBOTHRYS NOTHOFULVUS Gray. Frequent on open hillslopes across range.

PLAGIOBOTHRYS INFECTIVUS Johnston, Journ. Arn. Arb. **20**:380. 1939, type from lower Hospital Canyon, Mount Hamilton Range, *Hoover 3067* (G), isotype (UC!); Corral Hollow, *Hoover 1744* (UC).

VERBENACEAE

VERBENA PROSTRATA R. Br. Infrequent in dry ground of valley flats, interior of range.

LABIATAE

TRICHOSTEMA LANCEOLATUM Benth. Frequent on dry, open slopes, west side and interior of range.

SCUTELLARIA SIPHOCAMPYLOIDES Vatke. (*S. angustifolia* Pursh var. *canescens* Gray.) Occasional in rocky areas at stream margins, or flood beds of streams, interior of range.

SCUTELLARIA TUBEROSA Benth. Open hillsides or chaparral, interior and east side of range, infrequent.

*MARRUBIUM VULGARE L. Common only on neglected agricultural lands, west side of range near base.

SALVIA CARDUACEA Benth. Corral Hollow, *Brewer 1210* (UC), *Carter 787* (UC). Typically a desert species of southern California.

SALVIA COLUMBARIAE Benth. Dry, often rocky hillsides, interior and east side of range, abundant in scattered colonies.

SALVIA MELLIFERA Greene. East side of range as an occasional chaparral constituent, or sometimes abundant with *Artemisia californica*.

ACANTHOMINA LANCEOLATA Curran. Type from Calaveras Valley, Mount Hamilton Range, June 1878, *E. Brooks;* this area is now inundated by the Calaveras Reservoir. Rocky slopes, usually of loose shale across range, but mainly restricted to interior and east side. Mount Hamilton, *Greene* in 1891 (UC), *C. W. and H. K. Sharsmith 1233* (UC); Santa Isabella Valley, *C. W. and H. K. Sharsmith 1934* (UC); San Antonio Creek, *C. W. and H. K. Sharsmith 3194* (UC); Arroyo del Puerto, *C. W. and H. K. Sharsmith 3112* (UC), *Hoover 2628* (UC); Soda Springs Canyon, Pine Ridge, *Dudley 4151* (UC).

The genus *Acanthomintha* consists of three species, *A. ilicifolia* Gray, *A. lanceolata* Curran, and *A. obovata* Jepson, all restricted to California. In order to determine the position of *A. lanceolata*, a careful morphological study of the three species was made. They show close relationship and are similar in habit, but there is good evidence of specific differentiation, and intergradation is relatively minor or lacking.

Differences in bracts and leaves have been used to distinguish the three species, but these characters are not satisfactory for this purpose. The shape of the bracts is nearly uniform in all three species, and although the marginal spines vary somewhat in number they are most typically seven or nine throughout. The leaves tend to be very similar in all three species. Specific differences of a more reliable nature are tabulated below:

	A. lanceolata	A. obovata	A. ilicifolia
Herbage	Pilose puberulent, somewhat glandular.	Canescent or subglabrous.	Subglabrous or glabrous.
Calyx	Aristate teeth of upper lip 5-7 mm. long.	Aristate teeth of upper lip 1-3 mm. long.	Aristate teeth of upper lip 1-1.5 mm. long.
Corolla	Upper lip 2 - lobed at apex, 5 mm. long; lower lip 5-6 mm. long, median lobe linear.	Upper lip entire, 3 - 4 mm. long; lower lip 5-6 mm. long, median lobe of lower lip not elongate, lateral lobes broad.	Upper lip entire, 3-4 mm. long; lower lip 5-6 mm. long; median lobe not elongate, lateral lobes broad.
Anthers	Glabrous or sparsely hairy.	Woolly.	Glabrous.
Style	Sparsely hairy.	Glabrous.	Glabrous.

Of these differentiating characteristics, one that is not found in any of the species descriptions is that of the calyx teeth. On the basis of this and other diagnostic features, the following key to the genus is presented:

Aristate teeth of upper calyx lip 5-7 mm. long; anthers glabrous or sparsely hairy;
     style sparsely hairy ....*A. lanceolata*

Aristate teeth of upper calyx lip 1-3 mm. long; style glabrous.
     Anthers woolly ....*A. obovata*
     Anthers glabrous ....*A. ilicifolia*

Acanthomintha lanceolata is found in the Mount Hamilton Range and south in the San Carlos Range to the area where Gavilan and San Carlos ranges join (southern border of San Benito County and northern edge of Monterey County). At its northern boundary, *A. obovata* overlaps the southern boundary of *A. lanceolata* and extends south in the South Coast Ranges to the Mount Pinos region of Ventura County. *Acanthomintha ilicifolia* is usually considered as limited to the seacoast mesas of western San Diego County, but several collections have been made from San Mateo County in the outer South Coast Ranges which appear definitely to belong to *A. ilicifolia*, thus giving this species a very discontinuous distribution.

T. S. Brandegee (Zöe **4**:156. 1893) suggests that *Acanthomintha lanceolata* merges with *A. ilicifolia* in the southern San Carlos Range-Gavilan Range area. At least in part, Brandegee was apparently dealing with material of the then undescribed *A. obovata*. With the additional knowledge now available of the distribution and morphology of these species, his arguments do not seem tenable.

Acanthomintha obovata and *A. ilicifolia* have in common corolla shape and size and a glabrous style, and they may intergrade as to the subglabrous condition and the length of the aristate upper calyx teeth. *Acanthomintha lanceolata* is the most strongly differentiated species, being marked not only by differences in corolla shape and size, and sparsely hairy style, but as well by the pilose, somewhat glandular pubescence, and the very long aristate upper calyx teeth. *Acanthomintha ilicifolia* appears to be the most primitive species, and *A. lanceolata* the most advanced. Analysis of morphology along with geographic distribution suggests that *A. ilicifolia* may once have occupied more or less continuously the coastal region of California between San Diego County and central California. *Acanthomintha obovata* and *A. lanceolata* appear to have arisen directly or indirectly as offshoots of this species, replacing it in much of its original range, so that *A. ilicifolia* is now restricted to the northern and southern extremities of the area occupied by the genus as a whole.

Pogogyne serpylloides (Torr.) Gray. Occasional in chaparral, interior of range.

Satureja Douglasii (Benth.) Briq. (*Micromeria Chamissonis* (Benth.) Greene.)

Wooded areas, west side of range, infrequent. Typically an outer Coast Range species; seldom collected from inner Coast Ranges.

STACHYS PYCNANTHA Benth. Near springs, Mount Hamilton.

STACHYS AJUGOIDES Benth. Moist depressions, Santa Isabella Valley.

STACHYS RIGIDA Nutt. subsp. QUERCETORUM (Heller) Epling. (*S. quercetorum* Heller; *S. bullata* of authors, non Benth.) Occasional on wooded slopes, west side and interior of range.

*LAMIUM AMPLEXICAULE L. Edge of stream, Arroyo Bayo.

MONARDELLA VILLOSA Benth. subsp. SUBSERRATA (Greene) Epling. (*M. villosa* var. *tomentosa* Jepson.) Frequent on rocky slopes across range.

MONARDELLA DOUGLASII Benth. Dry, rocky slopes, occasional across range.

MONARDELLA BREWERI Gray, type from Corral Hollow, Mount Hamilton Range, *Brewer 1263*, June 3, 1892, isotype UC! The type locality represents the northern limit of distribution for this species; it extends south in the inner South Coast Ranges to the Mohave Desert.

SOLANACEAE

DATURA METELOIDES DC. Flood bed, Arroyo del Puerto.

PETUNIA PARVIFLORA Juss. Flood bed Santa Isabella Creek.

*NICOTIANA GLAUCA Graham. Canyon bottoms, east side of range.

NICOTIANA ATTENUATA (Torr.) Wats. Occasional in valleys or flood beds of streams, interior of range.

SOLANUM UMBELLIFERUM Esch. Occasional across range on openly wooded slopes.

SCROPHULARIACEAE

*VERBASCUM THAPSUS L. Arroyo del Vallé.

ANTIRRHINUM GLANDULOSUM Lindl. Infrequent in chaparral across range.

ANTIRRHINUM VEXILLO-CALYCULATUM Kell. var. TYPICUM Munz. (*A. vagans* Gray; *A. vagans* var. *rimorum* Jepson, type from Morrison Canyon, Mount Hamilton Range, *Jepson*.) Occasional on rocky slopes or in chaparral across range.

ANTIRRHINUM VEXILLO-CALYCULATUM Kell. var. BREWERI (Gray) Munz. (*A. vagans* var. *Breweri* Jepson.) North side Arroyo Bayo, *Mason 11699* (UC). Not reported hitherto from the South Coast Ranges. Although stems of the plants from the above collection are not "quite glandular pubescent throughout" as described by Munz, the plants otherwise match description and specimens of variety *Breweri*, and are sharply distinguished from the preceding variety *typicum*.

LINARIA TEXANA Scheele. (*L. canadensis* Dum.; *L. canadensis* var. *texana* Pennell.) Chaparral, Santa Isabella Valley.

COLLLINSIA HETEROPHYLLA Graham. (*C. bicolor* Benth., non Raf.) Frequent on wooded slopes across range. In the Mount Hamilton Range material the corolla varies widely and the calyces may be villous or glabrous within the same colony or between separate colonies. The length of the filament appendages in separating this and the other species of *Collinsia* as used by Newsom (Bot. Gaz. **87**:260-301. 1929) is an obscure basis for differentiation, and in the Mount Hamilton Range material is not reliable.

COLLINSIA BARTSIAEFOLIA Benth. Seeboy Ridge. Separated from *C. heterophylla* with considerable difficulty; *C. bartsiaefolia* has smaller flowers, narrower bases to the crenate leaves, and no pubescent appendages at the bases of the upper filaments.

COLLINSIA SPARSIFLORA Fisch. and Mey. var. SOLITARIA (Kellogg) Newsom. Frequent on wooded slopes across range.

TONELLA TENELLA (Benth.) Heller. Wooded slopes, Mount Hamilton.

SCROPHULARIA CALIFORNICA Cham. Occasional on wooded slopes, west side of range.

PENSTEMON CORYMBOSUS Benth. Rocky outcrops, west side of range.

PENSTEMON BREVIFLORUS Lindl. subsp. TYPICUS Keck. Occasional on chaparral slopes across range.

PENSTEMON HETEROPHYLLUS Lindl. subsp. TYPICUS Keck. Occasional on chaparral slopes across range.

PENSTEMON HETEROPHYLLUS Lindl. subsp. PURDYI Keck, type from Mount Hamilton, April, 1903, *Elmer 4832*. (*P. azureus* of authors, non Benth.) Open, rocky hillsides, west side of range.

PEDICULARIS DENSIFLORA Benth. Wooded areas, west side and interior of range, infrequent.

MIMULUS AURANTIACUS Curt. (*Diplacus aurantiacus* (Curt.) Jepson; *Diplacus glutinosus* Nutt.) Occasional on rocky slopes across range.

MIMULUS ANDROSACEUS Curran. (*M. Palmeri* Gray var. *androsaceus* (Curran) Gray.) Infrequent in chaparral, interior of range. Arroyo Bayo, *H. K. Sharsmith 1703* (UC); Red Mountains, *H. K. Sharsmith 3617* (UC). These two collections represent a long extension of known range for *M. androsaceus*, as this little known species has been considered to be restricted to the Tehachapi Mountains. There is a previous record of it from the South Coast Ranges, however, which has passed unnoticed. Elmer (Bot. Gaz. **41**:324. 1906), under *Eunanus androsaceus* Curran, makes the following comments: "From the middle western part of the state it is only known at Ben Lomond, Santa Cruz County, where fruiting specimens were collected by Mrs. K. Brandegee in April, 1890. In July, 1903, the writer found excellent flowering specimens in the same locality, which were distributed under *4519*. It is evidently rare and prefers hot and dry gravelly soil of the chaparral. . . . ." These two specimens cannot now be located.

MIMULUS ANDROSACEUS is closely allied to *M. Palmeri*, but is distinguished by the spreading, longer pedicels, the glabrous, truncate and mucronate calyx teeth, the smaller corolla with equal, entire lobes, and the glabrous anthers.

MIMULUS BOLANDERI Gray var. BRACHYDONTUS Grant (as *Eunanus Bolanderi* Greene, Erythea **1**:95. 1893.) Rocky hillsides or in chaparral, occasional across range.

MIMULUS FLORIBUNDUS Dougl. Moist flood bed Santa Isabella Creek.

MIMULUS CARDINALIS Dougl. Santa Isabella Creek.

MIMULUS NASUTUS Greene. (*M. guttatus* DC. var. *nasutus* Jepson; *M. guttatus* of authors in part.) Frequent in moist areas across range.

MIMETANTHE PILOSA (Benth.) Greene. Occasional in moist areas, interior of range.

VERONICA AMERICANA (Raf.) *Schweinitz*. Springs, north side Mount Hamilton.

VERONICA PEREGRINA L. subsp. XALAPENSIS (HBK) Pennell. Occasional in moist areas, interior of range.

CASTILLEJA ROSEANA Eastw. Rocky ridge, Copernicus Peak, *H. K. Sharsmith 920* (UC), *Morrison and Carter 3100* (UC). This species was based on the three existing specimens from the San Carlos Range (Eastwood, Leafl. West. Bot. **2**:104. 1938). The type was collected between San Lucas and Priest Valley, Monterey County (San Carlos Range), *Eastwood and Howell 2460* (CA, isotype CA!). The Mount Hamilton locality extends the known range of the species as follows: San Carlos Range and Mount Hamilton Range of inner South Coast Ranges.

Close morphological relationship is obvious between *Castilleja Roseana* and *C. latifolia* Hook. and Arn., the latter a species of the immediate seacoast between Monterey and Mendocino counties. In order to determine if *C. Roseana* justified specific distinction. from *C. latifolia*, fresh and herbarium specimens of both species were studied.

Castilleja latifolia shows a very considerable range of variation: the herbage varies from slightly glandular to viscid (in the variety *Wightii* Ziehle); the leaves range from 0.5 cm. long, entire and oval to 5 cm. long, ovate-lanceolate, and with one or two pairs of lateral lobes; the bracts vary from yellow to scarlet or crimson, and from 1 cm. long, entire and oval to 2.5 cm. long, triangular-obovate, and with one or two pairs of weakly or well developed lobes, the central segment rounded to erose-truncate at apex; the calyx is 12-30 mm. long, with sagittal incisions 6-15 mm. deep, the lateral lobes scarcely evident to well developed; the corolla is 15-30 mm. long, the galea 8-15 mm. long, with the lower lip either included or somewhat exserted.

Castilleja Roseana could fit into the extremes of variation represented above for *C. latifolia*, except that the plants are very viscid glandular throughout (even much more so than in *C. latifolia* var. *Wightii*), and the leaves are strongly undulant margined ("attenuate and crisped" according to the original description). The leaves are striking even in the herbarium mounts, for the leaf margins appear erose when flattened, although this is not so evident when the plants are fully mature. Due to the usual presence of a pair of divaricate lobes, the bracts are typically broader than long in C. *Roseana*, but this same shape is found occasionally among the multishaped bracts of C. *latifolia*. Certain minor differences were noted in the fresh flowers of C. *Roseana* as compared to the fresh flowers of C. *latifolia*: the galea was diffused with red throughout and the style pink in C. *Roseana*, while the galea was red only along the margins and the style green in C. *latifolia*. Whether or not these distinctions hold throughout the range of C. *latifolia* cannot be determined accurately from herbarium material, as this species often loses color in the pressed plants.

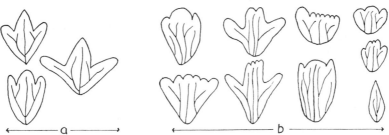

Fig. 10. Variations in bracts. a. *Castilleja Roseana*, ×2/3. b. *C. latifolia*, ×2/3.

From the above, it may seem that the morphological bases for distinguishing C. *Roseana* from C. *latifolia* are meager, but C. *Roseana* is relatively constant in its character and there is no significant variation between San Carlos Range and Mount Hamilton material of this species. When geographical distinction is added to morphological, the specific recognition of C. *Roseana* seems justified. Geographically, and ecologically as well, the inner Coast Range C. *Roseana* is strongly isolated from the strictly maritime C. *latifolia* which occurs only along a very narrow strip of seacoast.

CASTILLEJA FOLIOLOSA Hook. and Arn. Frequent on dry, open hills across range.

CASTILLEJA DOUGLASII Benth. (*C. parviflora* Bong. var. *Douglasii* Jepson.) Occasional across range on wooded or grassy slopes.

CASTILLEJA AFFINIS Hook. and Arn. Occasional on openly wooded slopes across range. *H. K. Sharsmith 1979* (UC) from Seeboy Ridge has light salmon pink bracts, and is dubiously referred to this species. Much intergradation occurs between C. *affinis* and C. *Douglasii*, and their specific distinction is questioned.

ORTHOCARPUS PUSILLUS Benth. var. TYPICUS Keck. Occasional on openly wooded slopes across range.

ORTHOCARPUS ERIANTHUS Benth. var. TYPICUS Keck. Occasional; openly wooded slopes across range.

ORTHOCARPUS ERIANTHUS Benth. var. MICRANTHUS (Gray) Jepson. Grassland, Arroyo del Puerto.

ORTHOCARPUS ATTENUATUS Gray. Occasional in grassland, interior and east side of range.

ORTHOCARPUS DENSIFLORUS Benth. var. TYPICUS Keck. Occasional on openly wooded slopes, west side and interior of range.

ORTHOCARPUS PURPURASCENS Benth. var. TYPICUS Keck. Frequent vernal annual in valleys or on grassy slopes across range.

CORDYLANTHUS RIGIDUS (Benth.) Jepson var. BREVIBRACTEATUS (Gray) Jepson. (*Adenostegia rigida* var. *brevibracteata* Greene.) Occasional autumnal species on openly wooded slopes, interior and east side of range.

OROBANCHACEAE

OROBANCHE UNIFLORA L. var. SEDI (Suksd.) Achey. Infrequent on rocky, wooded slopes, interior and east side of range.

OROBANCHE FASCICULATA Nutt. (*Aphyllon fasciculatum* Gray.) Parasitic on *Eriogonum* spp., *Artemisia* spp., etc., occasional in interior and east side of range.

PLANTAGINACEAE

*PLANTAGO MAJOR L. Moist sand, Santa Isabella Creek.

PLANTAGO HOOKERIANA Fisch. and Mey. var. CALIFORNICA (Greene) Poe. (*P. erecta* Morris.) Frequent vernal annual in grassland across range.

RUBIACEAE

GALIUM APARINE L. Wooded hillsides, west side and interior of range.

GALIUM ANDREWSII Gray. Rocky outcrop, Santa Isabella Creek.

GALIUM NUTTALLI| Gray. Wooded slopes, infrequent across range.

CAPRIFOLIACEAE

SAMBUCUS COERULEA Raf. (*S. glauca* Nutt.) Occasional in canyons, west side and interior of range.

SYMPHORICARPUS ALBUS (L.) Blake. Common on west side of range, forming thickets on north-facing slopes.

SYMPHORICARPOS MOLLIS Nutt. (*S. albus* var. *mollis* (Nutt.) Keck.) Low compact shrub in wooded slopes, interior of range.

LONICERA JOHNSTONI (Keck) McMinn. (*L. subspicata* var. *Johnstonii* Keck; *L. subspicata* of authors in part.) Frequent on brushy slopes across range. Predominately of southern California, although found in the inner South Coast Ranges to Mount Diablo.

LONICERA INTERRUPTA Benth. Occasional across range on brushy slopes. Some intergradation occurs in the Mount Hamilton Range between the mainly southern *L. Johnstoni* and the more northern *L. interrupta* where the two species overlap.

LONICERA HISPIDULA Dougl. (*L. hispidula* var. *californica* Greene; *L. hispidula* var. *vacillans* Gray.) Mount Hamilton; wooded areas.

VALERIANACEAE

PLECTRITIS SAMOLIFOLIA (DC.) Hoeck. Occasional in moist areas, interior of range.

PLECTRITIS MACROCERA Torr. and Gray. Frequent vernal annual on wooded slopes, west side and interior of range.

PLECTRITIS MAGNA (Greene) Suksd. Occasional in grassy areas across range.

PLECTRITIS CILIOSA (Greene) Jepson. Frequent vernal annual on wooded slopes across range.

ALIGERA RUBENS Suksd. Wooded hills, Arroyo Bayo.

ALIGERA COLLINA (Heller) Suksd. (*Plectritis collina* Heller (Muhl. **2**:329. 1907), type from western slope Copernicus Peak, Mount Hamilton Range, *Heller 8609*). Isotype WSC! Apparently a localized species; recollected on Mount Hamilton in 1938 by Dyal, but her specimen not seen by the writer.

DIPSACACEAE

*DIPSACUS FULLONUM L. Springs, north side Mount Hamilton.

CUCURBITACEAE

ECHINOCYSTIS FABACEA Naud. (*Micrampelis fabacea* Greene.) Occasional in wooded areas across range.

CAMPANULACEAE

CAMPANULA EXIGUA Rattan. In isolated colonies across range, mainly on unstable talus of chaparral belt. *Campanula exigua* is restricted to Mount Diablo, Mount Hamilton Range, and San Carlos Range, all of the inner South Coast Ranges. The closely related species, *C. angustiflora* Eastw., is restricted to Mount Tamalpais, Cobb Mountain, Howell Mountain, and Mount St. Helena of the North Coast Ranges, with a recently described variety, *C. angustiflora* var. *exilis* Howell (Leafl. West. Bot. **2**:102. 1938) occurring in the Pinnacles of San Benito County.

HETEROCODON RARIFLORUM Nutt. Moist areas, interior of range.

GITHOPSIS SPECULARIOIDES Nutt. Occasional in rocky areas, west side and interior of range.

LOBELIACEAE

NEMACLADUS MONTANUS Greene. (*Nemacladus rigidus* Curran var. *montanus* (Greene) Munz.) Unstable talus, Red Mountains.

NEMACLADUS RAMOSISSIMUM Nutt. Infrequent on rocky slopes, interior and east side of range. Possibly cne of the varietal forms of the species.

COMPOSITAE

MICROSERIS ACUMINATA Greene. (*M. Douglasii* and *M. Bigelovii* of authors.) Adobe Valley, *H. K. Sharsmith 3558* (UC). Known from the North Coast Ranges, Napa County to Humboldt and Tehama counties, and from the Sierra Nevada foothills in Eldorado County. Many of the collections of *M. acuminata* are found in herbaria under *M. Douglasii* and *M. Bigelovii*. The latter occurs in the South Coast Ranges as well as northward, and it is probable that many of the *M. Bigelovii* South Coast Range specimens are actually *M. acuminata*. Among a group of species wherein the lines of differentiation are very difficult to draw, definite conclusions cannot be offered without detailed study.

MICROSERIS TENELLA (Gray) Sch. Bip. var. APHANTOCARPHA (Gray) Black. (*M. elegans* Greene.) Frequent in grassy areas, interior and east side of range.

MICROSERIS DOUGLASII (DC). Sch. Bip. Frequent on grassy slopes across range.

MICROSERIS LINEARIFOLIA (DC.) Sch. Bip. (*Calais linearifolia* DC.; *Uropappus linearifolius* Nutt.) Common on open slopes across the range.

MICROSERIS LINDLEYI (DC.) Gray. (*Uropappus Lindleyi* DC.) Arroyo Mocho. Much less frequent than preceding species, but covering the same general habitats and area.

MICROSERIS SYLVATICA (Benth.) Sch. Bip. (*Scorzonella sylvatica* Benth.) Occasional on open slopes, interior and east side of range.

RAFINESQUIA CALIFORNICA Nutt. (*Nemoseris californica* Greene.) Wooded hillsides across range.

*HYPOCHOERIS GLABRA L. Mount Hamilton.

STEPHANOMERIA VIRGATA Benth. (*Ptiloria canescens* Greene.) Dry canyon slopes, occasional across range.

STEPHANOMERIA EXIGUA Nutt. var. CORONARIA (Greene) Jepson. Dry hillsides, interior of range; infrequent.

*LACTUCA SALIGNA L. Adobe Creek, Red Mountains.

Sonchus asper L. Infrequent, moist areas, interior of the range.

MALOCOTHRIX COULTERI Harv. and Gray. Infrequent on dry, grassy slopes, east margin of range. Mainly desert and cismontane, reaching its northern outpost at Antioch, Contra Costa County.

MALOCOTHRIX CLEVELANDII Gray. Occasional in chaparral, interior of range.

MALOCOTHRIX FLOCCIFERA (DC.) Blake. (*M. obtusa* Benth.) Frequent on serpentine talus, Red Mountains and east slope of Mount Day Ridge, occasional on non-serpentine areas, interior of range.

AGOSERIS GRANDIFLORA (Nutt.) Greene. Occasional on rocky, wooded slopes, west side and interior of range.

AGOSERIS HETEROPHYLLA (Nutt.) Greene var. CALIFORNICA (Nutt.) Jepson. Occasional to frequent on grassy slopes or valley flats, interior of range.

AGOSERIS HETEROPHYLLA var. KYMAPLEURA Greene. Frequent in grasslands across range.

AGOSERIS PLEBEIA Greene. Wooded slopes, west side and interior of range.

HIERACIUM ALBIFLORUM Hook. Rocky slope, Mount Hamilton.

CREPIS OCCIDENTALIS Nutt. subsp. PUMILA (Rydb.) Babc. and Stebbins. (*C. occidentalis* of authors, non Nutt.) Mount Hamilton, *Elmer 4872* (UC); Colorado Creek, *H. K. Sharsmith 3185* (UC). "Mount Hamilton, *Brewer 1304*" according to Coville, Contrib. U. S. Nat. Herb. 3:561. 1896. These Mount Hamilton Range localities are the only South Coast Range areas known for this subspecies, which is described as apomict *hamiltonensis* by Babcock and Stebbins (Carn. Inst. Wash. Publ. **504**:128, 132. 1938).

CREPIS MONTICOLA Cov. Wooded area, summit of Seeboy Ridge, *H. K. Sharsmith 3059* (UC). Typically of North Coast Ranges, occurring south of Lake County only in Mount Hamilton Range, the Mount Hamilton specimens being designated by Babcock and Stebbins *(op. cit)* as apomict *australis.*

GUTIERREZIA CALIFORNICA (DC.) Torr. and Gray. Dry, rocky areas, interior and east side of range.

GRINDELIA CAMPORUM Greene var. DAVYI (Jepson) Steyermark. (*G. robusta* Nutt. var. *Davyi* Jepson.) Occasional in rocky areas, interior and east side of range.

GRINDELIA CAMPORUM Greene var. PARVIFLORA Steyermark. Occasional; dry slopes, interior and east side of range.

GRINDELIA RUBRICAULIS DC. Occasional on exposed slopes, west side and interior of range. Questionably distinct from *G. camporum* var. *interioris.*

STENOTOPSIS LINEARIFOLIUM (DC.) Rydb. (*Haplopappus linearifolius* DC.) Chaparral element, interior and east side of range.

ERICAMERIA ARBORESCENS (Gray) Greene. (*Haplopappus arborescens* (Gray) Hall.) Chaparral areas, interior of range.

EASTWOODIA ELEGANS Brandegee. Hillsides, east side of range. Restricted to the extreme eastern side of inner South Coast Ranges from Corral Hollow to Maricopa Hills in Kern County.

CHRYSOTHAMNUS NAUSEOSUS (Pallas) Brit. var. MOHAVENSIS (Greene) Hall. (*Bigelovia mohavensis* Greene; *C. mohavensis* Greene.) Summit rocks, Copernicus Peak, *H. K. Sharsmith 1404, 1418* (UC); loose talus, Colorado Creek, Red Mountains, *H. K. Sharsmith 3892* (UC). A well defined variety in a highly polymorphic species; typically of western borders of Mohave Desert, found infrequently northward in the inner South Coast Ranges to Mount Hamilton Range. Only two other stations are known in the South Coast Ranges, both in San Carlos Range. Greene first found it upon Mount Hamilton in 1893 (Erythea **1**:89. 1893).

SOLIDAGO CALIFORNICA Nutt. Occasional in moist gullies across range.

CHRYSOPSIS VILLOSA (Pursh.) Nutt. var. ECHIOIDES (Benth.) Gray. (*C. echioides* Benth.) Exposed, dry slopes, Mount Hamilton, *H. K. Sharsmith 1354, 1411* (UC). Much variation occurs, and more than variety *echioides* may be represented in the collections cited, but this highly polymorphic genus has been only poorly studied.

CHRYSOPSIS VILLOSA (Pursh.) Nutt. var. SESSILIFLORA (Nutt.) Gray. Grassland, western slopes of range.

CHRYSOPSIS OREGANA (Nutt.) Gray var. SCABERRIMA Gray. Occasional in dry flood beds, interior of range.

PENTACHAETA EXILIS Gray. (*P. exilis* var. *aphanochaeta* Gray.) Occasional on grassy slopes across range.

PENTACHAETA LAXA Elmer, Bot. Gaz. **41**:318. 1906, is based upon a collection from Cedar Mountain, Mount Hamilton Range, May, 1903, *Elmer 4437.* According to Elmer, "This distinct species..... evidently very rare." Examination of the isotype of *P. laxa* (UC!) and analysis of the type description proves the species to be *Baeria microglossa* (Gray) Greene of the tribe *Helenieae.*

LESSINGIA HOLOLEUCA Greene. (*L. leptoclada* var. *hololeuca* Jepson.) Grassy slopes, west side of range.

LESSINGIA GERMANORUM Cham. var. PARVULA (Greene) Howell. (*L. parvula* Greene; *L. tenuis* var. *Jaredii* Jepson.) Frequent chaparral associate in interior of range. A highly polymorphic species represented by ten variants, all of which show intergradation where they overlap. In general, however, the Mount Hamilton Range material represents the variety *parvula* (J. T. Howell, Univ. Calif. Publ. Bot. **16**: 17. 1929).

LESSINGIA NEMACLADA Greene. (*L. ramulosa* var. *microcephala* Jepson.) Open chaparral, rocky slopes, Red Mountains. Arroyo del Puerto, *H. K. Sharsmith 3789* (UC); San Antonio Valley, *H. K. Sharsmith 3906* (UC). These collections represent the species typical more nearly than the widespread var. *mendocino* (Greene) Howell or the little known var. *albiflora* (Eastw.) Howell.

Lessingia nemaclada variety *albiflora* (Eastw.) Howell was collected near San Antonio Valley by D. D. Keck. The collection was not seen by the writer.

CORETHROGYNE FILAGINIFOLIA (Hook. and Arn.) Nutt. var. TYPICA Canby. Frequent on dry, exposed slopes, west side of range, occasional in interior.

ASTER MENZIESII Lindl. Dry hillsides, Mount Hamilton.

ERIGERON MISER Gray. Rocky ridges, west side of range.

ERIGERON PHILADELPHICUS L. Moist areas, west side and interior of range.

BACCHARIS PILULARIS DC. var. CONSANGUINEA Wolf. Occasional, west side of range in shallow ravines.

BACCHARIS DOUGLASII DC. Occasional, moist areas, interior of range.

BACCHARIS VIMINEA DC. Occasional along stream margins, interior and east side of range. *H. K. Sharsmith 3767* (UC) from Arroyo del Puerto approaches the leaf characteristics of *B. Douglasii*, but has the floral characteristics and shrubby habit of *B. viminea*.

MICROPUS CALIFORNICUS Fisch. and Mey. Occasional on exposed hillsides across range.

FILAGO CALIFORNICA Nutt. Occasional on exposed hillsides across range.

STYLOCLINE GNAPHALOIDES Nutt. Open chaparral, interior of range.

STYLOCLINE FILAGINEA Gray. Occasional on brushy or chaparral slopes, interior and east side of range.

PSILOCARPHUS TENELLUS Nutt. Occasional in moist areas of valleys across range.

EVAX SPARSIFLORA (Gray) Jepson (probably var. BREVIFOLIA (Gray) Jepson.) Rocky areas, interior of range.

GNAPHALIUM PALUSTRE Nutt. Stream flood beds or margins of vernal pools, west side and interior of range.

GNAPHALIUM CALIFORNICUM DC. (*G. decurrens* Ives var. *californicum* (DC.) Gray.) Open hillslopes, west side and interior of range. If *G. californicum* were to be considered as a variety of the Great Basin species known as *G. decurrens* Ives (intergradation does occur to some extent), a new combination would be necessary, for *G. decurrens* Ives, which preceded *G. californicum* as to date of publication, is a later homonym of *G. decurrens* L. Greene first recognized this (Ottawa Nat. **15**:278. 1902) and published the new binomial *G. Macounii* Greene.

GNAPHALIUM CHILENSE Spreng. Occasional in flood beds, interior and east side of range.

HELIANTHUS CALIFORNICUS DC. Stream bed, Adobe Creek.

BALSAMORHIZA MACROLEPIS Sharp. (*B. Hookeri* of authors, non Nutt.) Occasional, rocky slopes, west side and interior of range.

WYETHIA ANGUSTIFOLIA (DC.) Nutt. Rocky slope, west side Mount Hamilton.

WYETHIA HELENIOIDES (DC.) Nutt. Scattered, isolated plants common on west side and interior of range.

HELIANTHELLA CALIFORNICA Gray. Rocky wooded or chaparral areas, west side and interior of range. It is questionable if *H. californica* of the central Coast Ranges is distinct from *H. castanea* Greene of Mount Diablo; both appear to lack satisfactory geographical separation and morphological distinctions.

COREOPSIS HAMILTONII (Elmer) H. K. Sharsmith, Madroño **4**:214. 1938. (*Leptosyne hamiltonii* Elmer, Bot. Gaz. **41**: 323. 1906, type from Mount Hamilton, April, 1900, *Elmer 2328*) Topotypes: Copernicus Peak, *H. K. Sharsmith 914, 1839* (UC); Mount Hamilton, *Eastwood 11671* (CA). Other collections: Mount Hamilton-Livermore road, *Eastwood 12468* (CA); San Antonio Valley, *Wieser* (S); Arroyo Bayo, *H. K. Sharsmith 1709, 3628, 3489* (UC). *Coreopsis hamiltonii* is known only from exposed, dry rocky slopes of the west summits and interior of Mount Hamilton Range.

COREOPSIS CALLIOPSIDEA (DC.) Gray. (*Leptosyne calliopsidea* Gray.) Abundant in isolated colonies, hillsides of east side and interior of range. A typically desert species, occasional in the inner South Coast Ranges to Corral Hollow of Mount Hamilton Range.

COREOPSIS DOUGLASII (DC). Hall (as to name but not as to description). *Leptosyne Douglasii* DC.; *C. Stillmanii* var. *Jonesii* Sherff.) Dry, shale slopes, Arroyo Bayo,

H. K. Sharsmith 3490, 3627, 3944, 3946 (UC). *Coreopsis Douglasii* occurs on dry rocky slopes of the inner South Coast Ranges from the Mount Hamilton Range to San Luis Obispo County. The relationships of this species to *C. californica* Nutt. of southern California are discussed in a separate publication of the writer (Madroño 4:209-231. 1938.)

COREOPSIS STILLMANII (Gray) Blake. (*Leptosyne Stillmanii* Gray.) In localized colonies on rocky, dry slopes, interior and east side of range.

HEMIZONIA PUNGENS (Hook. and Arn.) Torr. and Gray subsp. INTERIOR Keck. (*Centromadia pungens* Greene.) Chaparral slopes, east side of range.

HEMIZONIA FITCHII Gray. (*Centromadia Fitchii* Greene.) Dry, exposed areas across range.

HEMIZONIA CONGESTA DC. subsp. LUZULAEFOLIA Babc. and Hall. Dry slopes, west side and interior of range.

HEMIZONIA FASCICULATA (DC.) Torr. and Gray. Hillslope, Arroyo Mocho.

HEMIZONIA VIRGATA Gray. Abundant and conspicuous element of aestival flora on dry hillsides across range.

HEMIZONIA HEERMANNII Greene. (*H. virgata* Gray var. *Heermannii* Jepson.) Growing in similar habitats to *H. virgata*, east side and interior of range.

HEMIZONIA KELLOGGII Greene. (*H. Wrightii* Gray var. *Kelloggii* Jepson.) Occasional in flood beds of streams, interior and east side of range.

HEMIZONIA OBCONICA Clausen and Keck, type from near Tesla, Mount Hamilton Range, *Keck and Stockwell 2501* (S). A species frequently confused with *H. virgata*.

HEMIZONIA PLUMOSA (Kell.) Gray (*Calycadenia plumosa* Kell.; *Blepharizonia plumosa* Greene; *B. laxa* Greene.) Rocky dry slopes, east side of range.

CALYCADENIA VILLOSA DC. (*Hemizonia villosa* (DC.) Jepson.) Colorado Creek, *Keck 2487* (S).

CALYCADENIA TRUNCATA DC. (*Hemizonia truncata* Gray.) Dry, exposed areas, frequent in aestival flora across range.

CALYCADENIA HISPIDA Greene. Dry areas, interior and east side of range. Santa Isabella Valley, *C. W. and H. K. Sharsmith 1369* (UC); San Antonio Valley, *C. W. and H. K. Sharsmith 1379* (UC). The typical phase of the species is not represented in the Mount Hamilton Range; these collections should probably be referred to a subspecies.

CALYCADENIA MULTIGLANDULOSA DC. (*Hemizonia multiglandulosa* Gray.) Occasional on dry slopes, interior of range.

CALYCADENIA CEPHALOTES (Gray) Greene. (*Hemizonia multiglandulosa* var. *cephalotes* Gray; *H. cephalotes* Greene; *Calycadenia multiglandulosa* var. *cephalotes* Jepson.) Grassy, open slopes, Mount Hamilton.

MADIA MADIOIDES (Nutt.) Greene. West side of range on wooded slopes, infrequent.

MADIA ELEGANS Don. subsp. TYPICA Keck. (*M. elegans* var. *hispida* Hall.) Occasional across range on dry, rocky slopes.

MADIA RADIATA Kellogg. Hospital Canyon, *Chamberlin 6196* (S), *Hoover 3065* (UC). A distinctive species which occurs in isolated localities on the eastern margin of the inner South Coast Ranges from Antioch, Contra Costa County, to San Luis Obispo County.

MADIA SATIVA Mol. subsp. DISSITIFLORA (Nutt.) Keck. (*M. dissitiflora* Torr. and Gray.) Dry slopes, west side and interior of range.

MADIA EXIGUA (Smith) Gray. (*Harpaecarpus exiguus* Gray.) Occasional on exposed slopes across range.

LAYIA CHRYSANTHEMOIDES (DC). Gray. Very abundant in Santa Isabella Valley, forming extensive pure colonies or mixed with *L. platyglossa*.

LAYIA PLATYGLOSSA (Fisch. and Mey.) Gray. (*Blepharipappus platyglossus* Greene.) Frequent vernal annual, valley flats across the range.

LAYIA GAILLARDIOIDES (Hook. and Arn.) DC. Frequent on grassy or rocky slopes across range.

LAYIA HIERACIOIDES (DC.) Hook. and Arn. In chaparral, interior of range.

HEMIZONELLA MINIMA Gray. Occasional in chaparral, interior of range. Arroyo Bayo, *H. K. Sharsmith 1705* (UC); Burnt Hills, *H. K. Sharsmith 3457;* Copernicus

Peak, *H. K. Sharsmith 3712* (UC). The first records of this species in the South Coast Ranges.

LAGOPHYLLA RAMOSISSIMA Nutt. subsp. TYPICA Keck. Frequent aestival annual across range in exposed regions.

LAGOPHYLLA RAMOSISSIMA Nutt. subsp. CONGESTA (Greene) Keck. Infrequent, valley areas across range.

HOLOZONIA FILIPES (Hook. and Arn.) Greene. Dry streamlets, interior of the range. Santa Isabella Valley, *H. K. Sharsmith 3332* (UC); Arroyo Bayo, *H. K. Sharsmith 3866* (UC). Not before collected in the South Coast Ranges.

ACHYRACHAENA MOLLIS SCHAUER. Occasional on dry, open areas across the range.

*XANTHIUM SPINOSUM L. Abandoned farmyard, Mount Day Ridge.

LASTHENIA GLABRATA Lindl. Heavy adobe, Hall's Valley.

LASTHENIA GLABERRIMA DC. Vernal pool, Santa Isabella Creek.

BAERIA CHRYSOSTOMA Fisch. and Mey. Vernal annual, valley floors across range.

BAERIA CHRYSOSTOMA var. GRACILIS (DC.) Hall. (*B. gracilis* Gray.) Vernal annual, area of preceding but more common.

BAERIA ULIGINOSA (Nutt.) Gray. Valley flats, east side of range.

BAERIA MICROGLOSSA (DC.) Greene. Moist areas, interior and east side of range, infrequent.

MONOLOPIA MAJOR DC. Very abundant in isolated colonies, open hillsides, east side of range.

MONOLOPIA LANCEOLATA Nutt. Open hillsides, east side of range, frequently mingling with the above species.

ERIOPHYLLUM WALLACEI Gray. Several plants of this species were collected in the dry, gravelly stream bed near the headwaters of Arroyo Mocho, September 24, 1933, *C. W. and H. K. Sharsmith 441*. The plants were identified by L. Constance. Later they were lost, and none were found in subsequent seasons. The species is characteristic of the Colorado and Mohave deserts and the dryer interior valleys of southern California, extending into the Great Basin to the east. One specimen is in the University of California herbarium from Bakersfield, but otherwise the species appears not to have been collected outside of the areas mentioned above. Its infrequent occurrence in the inner South Coast Ranges north to the Mount Hamilton Range is not improbable, however, on the basis of the over forty other typically desert species which have been found to have this same distribution.

ERIOPHYLLUM CONFERTIFLORUM (DC.) Gray. Frequent shrub on rocky hillsides, interior of range.

ERIOPHYLLUM JEPSONII Greene, type from hills between Arroyo Mocho and Arroyo del Vallé, north end Mount Hamilton Range, May, 1891, *Jepson*. Frequent chaparral associate, interior and east side of range; restricted to Mount Diablo, Mount Hamilton Range, and San Carlos Range. In the area between Arroyo Bayo and Santa Isabella Valley where *E. Jepsonii* and *E. confertiflorum* occur together, typical plants of both species are found in the chaparral along with plants which share the characteristics of both, an indication that the two species hybridize.

ERIOPHYLLUM LANATUM (Pursh) Forbes var. ACHILLAEOIDES (DC.) Jepson. Arroyo Mocho, *Elmer 4335*, isotype (UC!) of *E. Greenei* Elmer, Bot. Gaz. 41:313. 1906.

RIGIOPAPPUS LEPTOCLADUS Gray. Frequent vernal annual on rocky slopes across range.

CHAENACTIS HETEROCARPHA Gray. (*Chaenactis glabriuscula* DC. var. *heterocarpha* (Gray) Hall.) Frequent in chaparral, interior and east side of range, often on serpentine.

HULSEA HETEROCHROMA Gray. Chaparral clearing, divide between Arroyo Mocho and Colorado Creek, *H. K. Sharsmith 959* (UC). Infrequent in the South Coast Ranges, known from isolated stations only.

HELENIUM PUBERULUM DC. Near streams, west side and interior of range; infrequent.

ACHILLEA LANULOSA Nutt. (*A. millefolium* L. var. *lanulosa* (Nutt.) Piper.) Frequent on dry slopes across range.

*MATRICARIA MATRICARIOIDES (Less.) Porter. (*M. suaveolens* (Pursh) Buch.; *M. discoidea* DC.) Common only near the cultivated margins of range.

ARTEMISIA CALIFORNICA Less. Frequent on exposed hillsides, west and east margins of range, not occurring in the interior.

ARTEMISIA VULGARIS L. var. CALIFORNICA Besser. (*A. vulgaris* var. *heterophylla* Jepson; *A. Douglasiana* Besser.) Occasional along streams across range.

ARTEMISIA DRACUNCULUS L. (*A. dracunculoides* of authors, non Pursh.) Occasional at stream margins, west side and interior of range.

CROCIDIUM MULTICAULIS Hook. Occasional on grassy slopes, interior of range. The Mount Hamilton Range is the only known locality in the South Coast Ranges for this northern species.

SENECIO DOUGLASII DC. Frequent shrub; exposed dry slopes across range.

SENECIO ARONICOIDES DC. Brush slopes or in chaparral, occasional on west side of range.

SENECIO BREWERI Davy. Infrequent on wooded slopes, interior and east side of range. Seeboy Ridge, *H. K. Sharsmith 1978* (UC); San Antonio Valley, *H. K. Sharsmith 3086* (UC); Adobe Valley, *H. K. Sharsmith 3559* (UC). A desert species occurring north in the inner South Coast Ranges to the Mount Hamilton Range, closely related to *S. eurycephalus* Torr. and Gray of the North Coast Ranges and considered as synonymous with this latter species by Hall (Univ. Calif. Publ. Bot. **3**:233. 1907). *Senecio Breweri* is quite glabrous, whereas *S. eurycephalus* is floccose or white tomentose, becoming glabrate with age. Hall considered these differences in pubescence to be related to habitat, the tomentose plants growing in less arid soil. This view, however, is not consistent with the geographic distribution of the two species and the localities occupied. *Senecio Breweri* is also distinguished from *S. eurycephalus* by more leaflets per leaf, more numerous heads, a corymbose inflorescence, and broad involucral bracts.

*SENECIO VULGARIS L. Occasional, lower western slopes of range.

ARNICA DISCOIDEA Benth. Wooded slopes, Mount Hamilton.

ARNICA CORDIFOLIA Hook. (*A. latifolia* of authors.) Wooded slopes, west side of range.

LEPIDOSPARTUM SQUAMATUM Gray. Dry streambed, Arroyo Mocho, *H. K. Sharsmith 444* (UC); Eylar Mountain, *Lundh 27* (VTM); Hospital Canyon, *Lundh 115* (VTM). Predominately a Great Basin and desert species; these are the first records of the species in the Mount Hamilton Range, apparently its northern limit of distribution.

CIRSIUM LANCEOLATUM (L.) Hill. Moist soil, Mount Day Ridge.

CIRSIUM CAMPYLON H. K. Sharsmith, Madroño **5**:85. 1939, type from Arroyo del Puerto, Red Mountains, Mount Hamilton Range, *H. K. Sharsmith 3761* (UC!), isotypes (G,K). *Cirsium campylon* is restricted to the Mount Hamilton Range. It forms dense but isolated colonies in moist, sandy soil along edges of small, perennial streams, all known localities occurring in areas of serpentine rock.

CIRSIUM CALIFORNICUM Gray. (*Carduus californicus* Greene.) Occasional on dry slopes, west side and interior of range. Smith Creek, *H. K. Sharsmith 1177* (UC); Grand View, *Pendleton 852* (UC); San Antonio Creek, *H. K. Sharsmith 3199* (UC); San Antonio Valley, *H. K. Sharsmith 3098* (UC). In the short peduncles and clustered heads, these specimens do not represent typical *C. californicus*. Collection *3199* is particularly atypical in these characteristics.

CIRSIUM OCCIDENTALE (Nutt.) Jepson var. COULTERI (Harv. and Gray) Jepson. Occasional on dry slopes across range.

*CENTAUREA MELITENSIS L. Cultivated areas, west base of range; also near a few abandoned homesites, interior of range.

Species to be expected in the Mount Hamilton Range

The following list consists of species which have been reported from the Mount Hamilton Range, but which have not been collected by the writer, nor have specimens been located in the herbaria consulted:

FESTUCA MYUROS L. Greene, Erythea **1**:97. 1893, Mount Hamilton.

ERIOGONUM HIRTIFLORUM Gray. Greene, *op. cit.*, p. 84, Mount Hamilton. This species reaches its best development in the chaparral areas of Lake County, but it has been recently reported from the Pinnacles of San Benito County in the South Coast Ranges (Howell, Leaf. West. Bot. **2**:99. 1938), and on the basis of this record and that of Greene, it is probable that it occurs in the Mount Hamilton Range.

LUPINUS NANUS Dougl. Heller, Muhl. **2**:292. 1907 (as *L. carnosulus* Greene), Alum Rock Park; Jepson, Fl. Calif. **2**:272. 1936 (as var. *apricus* C. P. Smith), Mount Hamilton.

EPILOBIUM CALIFORNICUM Hausskn. Greene, Erythea **1**:86. 1893, Joaquin [Murietta] Springs.

OSMORHIZA OCCIDENTALIS (Nutt.) Torr. Greene, *op. cit.*, p. 89 (as *Myrrhis occidentalis* (Nutt.) Benth. and Hook.), Mount Hamilton.

CICUTA DOUGLASII Coult. and Rose. Greene, *op. cit.*, p. 89 (as *Cicuta californica* Gray), Joaquin [Murietta] Springs.

CHIMAPHILA MENZIESII (R. Br.) Spreng. Greene, *op. cit.*, p. 92 (as *Pseva Menziesii* (R. Br.) O. Ktze.), Mount Hamilton.

APOCYNUM ANDROSAEMIFOLIUM L. Greene, *op. cit.*, p. 92, Mount Hamilton.

PHACELIA NEMORALIS Greene. Greene, *op. cit.*, p. 93 (as *P. circinata* (Willd.) Jacq.), Mount Hamilton.

OROBANCHE TUBEROSA (Gray) Heller. Greene, *op. cit.*, p. 95 (as *Aphyllon tuberosum* Gray), Mount Hamilton.

The following are introduced species which Greene reported from the summit of Mount Hamilton, but which have not been collected by the writer, nor have herbarium specimens been located:

POLYGONUM AVICULARE L. Greene, Erythea **1**:83. 1893.
CHENOPODIUM MURALE L. Greene, *op. cit.*, p. 85.
AMARANTHUS ALBUS L. Greene, *op. cit.*, p. 84.
AMARANTHUS RETROFLEXUS L. Greene, *op. cit.*, p. 84.
BRASSICA NIGRA (L.) Koch. Greene, *op. cit.*, p. 87.
MALVA PARVIFLORA L. Greene, *op. cit.*, p. 83.
SONCHUS OLERACEUS L. Greene, *op. cit.*, p. 92.
ANTHEMIS COTULA L. Greene, *op. cit.*, p. 91.
CENTAUREA SOLSTITIALIS L. Greene, *op. cit.*, p. 91.

DEPT. OF BOTANY,
UNIVERSITY OF MINNESOTA,
MINNEAPOLIS, MINN.

Index

This Index serves not only as a finding guide to the 761 species and varieties included in the *Flora of Mount Hamilton,* but also indicates name changes that have been adopted during the 36 years since the Flora was published. Names in italics are those used in Munz, *A California Flora and Supplement* (1973), which differ from those used in the *Flora of Mount Hamilton.* The usefulness of the Index is further enhanced by cross references from Munz back to Sharsmith.

The considerable work of compiling this Index was undertaken by Carl W. Sharsmith and Nobi Kurotori. The Santa Clara Chapter of the California Native Plant Society is deeply indebted to them and takes this opportunity to express thanks.

Acanthomintha lanceolata, 355
Acer macrophyllum, 346
Aceraceae, 346
Achillea lanulosa, 365
Achyrachaena mollis, 365
Adenostoma fasciculatum, 343
Adiantum Jordani, 328
Aesculus californica, 346
Agoseris
  grandiflora, 361
  heterophylla var californica, 361
    A. heterophylla
  heterophylla var kymapleura, 361
  plebeia, 361
    A. grandiflora
Agropyron subspicatum, 330
Agrostis verticillata, 330
  A. semiverticillata
Aizoaceae, 336
Alchemilla occidentalis, 343
Aligera
  collina, 360
    Plectritis macrocera
  rubens, 360

Plectritis ciliosa ssp insignis
Alisma Plantago-aquatica, 329
  A. triviale
Alismaceae, 329
  Alismataceae
Allium
  amplectens, 331
  Bolanderi, 332
  falcifolium, 332
  fimbriatum, 331
    A. fimbriatum var Sharsmithae
  lacunosum, 332
  Parryi, 331
    A. fimbriatum var diabolense
  peninsulare var crispum, 332
    A. crispum
  serratum, 331
  unifolium, 331
Allophyllum gilioides
  Gilia gilioides, 352
Alnus rhombifolia, 333
Alopecurus Howellii, 330
Amaranthaceae, 336
Amaranthus

albus, 367
blitoides, 336
A. graecizans
californicus, 336
retroflexus, 367
Amelanchier alnifolia, 343
A. pallida
Amsinckia
Eastwoodae, 354
A. intermedia var Eastwoodae
grandiflora, 354
intermedia, 354
parviflora, 354
A. Menziesii
tessellata, 354
vernicosa, 354
Anacardiaceae, 346
Anagallis arvensis, 350
Androsace occidentalis var acuta,
350
A. elongata var acuta
Angelica tomentosa, 349
Anthemis cotula, 367
Antirrhinum
glandulosum, 357
A. multiflorum
vexillo-calyculatum, 357
vexillo-calyculatum var Breweri,
357
A. Breweri
Apiastrum angustifolium, 349
Apocynaceae, 351
Apocynum
androsaemifolium, 367
cannabinum var glaberrimum,
351
Aquilegia
formosa ssp truncata, 338
Tracyi, 338
A. eximia
Arabis
Breweri, 341
glabra, 341
Aralia californica, 349
Araliaceae, 349
Arbutus Menziesii, 350
Arceuthobium campylopodum, 334

Arctostaphylos
glandulosa var Campbellae, 350
glauca, 350
Arenaria
californica, 337
Douglasii, 336
Douglasii var emarginata, 337
macrophylla, 336
pusilla, 337
Argemone platyceras, 340
A. munita
Arnica
cordifolia, 366
discoidea, 366
Artemisia
californica, 366
Dracunculus, 366
vulgaris var californica, 366
A. Douglasiana
Asclepiadaceae, 351
Asclepias
californica, 351
mexicana, 351
A. fascicularis
Aster Menziesii, 363
A. chilensis
Astragalus
didymocarpus, 345
Douglasii, 345
Gambelianus, 345
oxyphysus, 345
Athysanus pusillus, 341
Atriplex Serenana, 336
Avena
barbata, 330
fatua, 330
Baccharis
Douglasii, 363
pilularis var consanguinea, 363
viminea, 363
Baeria
chrysostoma, 365
Lasthenia chrysostoma
chrysostoma var gracilis, 365
Lasthenia chrysostoma ssp
gracilis
microglossa, 365

Lasthenia microglossa
uliginosa, 365
Lasthenia minor
Balsamorhiza macrolepis, 363
Barbarea orthoceras var
dolichocarpa, 341
Berberidaceae, 339
Berberis
dictyota, 339
pinnata, 339
Betulaceae, 333
Blepharizonia plumosa
Hemizonia plumosa, 364
Boisduvalia
densiflora, 348
stricta, 348
Boraginaceae, 354
Bowlesia incana, 349
Brassica
arvensis, 341
B. Kaber var pinnatifida
campestris, 341
incana, 341
B. geniculata
nigra, 367
Brodiaea
capitata, 332
B. pulchella
coronaria, 332
hyacinthina, 332
laxa, 332
peduncularis, 332
pulchella, 332
Bromus
anomalus, 329
arenarius, 329
breviaristatus, 329
carinatus, 329
mollis, 329
rigidus, 329
B. diandrus
rubens, 329
Calandrina ciliata var Menziesii,
336
Callitrichaceae, 346
Callitriche
marginata, 346

palustris, 346
C. verna
Calocedrus decurrens
Libocedrus decurrens, 329
Calochortus
albus, 332
clavatus, 332
C. clavatus ssp pallidus
invenustus, 332
luteus, 332
umbellatus, 332
venustus, 332
Calycadenia
cephalotes, 364
C. multiglandulosa ssp robusta
hispida, 364
multiglandulosa, 364
truncata, 364
villosa, 364
Calyptridium
monandrum, 336
Parryi, 336
Calystegia
malacophylla ssp pedicellata
Convolvulus malacophyllus,
351
subacaulis
Convolvulus californicus, 351
Campanula exigua, 360
Campanulaceae, 360
Caprifoliaceae, 360
Capsella
Bursa-pastoris, 341
procumbens, 341
Hutchinsia procumbens
Cardamine oligosperma, 341
Carex
densa, 331
nudata, 331
praegracilis, 331
serratodens, 331
Caryophyllaceae, 336
Castilleja
affinis, 359
Douglasii, 359
C. affinis
foliolosa, 359

Roseana, 358
Caucalis microcarpa, 349
Ceanothus
 cuneatus, 347
 Ferrisae, 347
 integerrimus, 347
 leucodermis, 347
 sorediatus, 347
Centaurea
 melitensis, 366
 solstitialis, 367
Centaurium floribundum, 350
Cerastium
 arvense var maximum, 336
  C. arvense
 viscosum, 336
  C. glomeratum
Cercocarpus betuloides, 343
Chaenactis heterocarpha, 365
  C. glabriuscula var
  megacephala
Chaetopappa exilis
 Pentachaeta exilis, 362
Cheilanthes
 gracillima, 328
 intertexta, 328
Chenopodiaceae, 336
Chenopodium
 album, 336
 californicum, 336
 murale, 367
Chimaphila Menziesii, 367
Chlorogalum pomeridianum, 331
Chorizanthe
 Clevelandii, 334
 membranacea, 334
 perfoliata, 335
 polygonoides, 335
 uniaristata, 334
Chrysopsis
 oregona var scaberrima, 362
 villosa var echioides, 362
 villosa var sessiliflora, 362
Chrysothamnus nauseosus var
 mohavensis, 362
Cicuta Douglasii, 367
Cirsium

californicum, 366
campylon, 366
lanceolatum, 366
  C. vulgare
occidentale var Coulteri, 366
  C. occidentale
Clarkia
 Breweri, 348
 concinna, 348
 elegans, 348
  C. unguiculata
 modesta
  Godetia epilobioides, 348
 purpurea
  Godetia quadrivulnera var
   Elmeri, 348
 purpurea ssp quadrivulnera
  Godetia quadrivulnera, 348
 rhomboidea, 348
 rubicunda
  Godetia amoena, 348
Claytonia
 gypsophiloides, 336
  Montia gypsophiloides
 perfoliata, 336
  Montia perfoliata
Clematis lasiantha, 339
Collinsia
 bartsiaefolia, 357
 heterophylla, 357
 sparsiflora var solitaria, 357
  C. sparsiflora var collina
Collomia
 diversifolia, 351
 grandiflora, 351
Compositae, 361
Conium maculatum, 349
Convolvulaceae, 351
Convolvulus
 californicus, 351
  Calystegia subacaulis
 malacophyllus, 351
  Calystegia malacophylla ssp
  pedicellata
Cordylanthus rigidus var
 brevibracteatus, 359

Coreopsis
  calliopsidea, 363
  Douglasii, 363
  Hamiltonii, 363
  Stillmanii, 364
Corethrogyne filaginifolia, 363
  *C. filaginifolia var
    hamiltonensis*
Cornaceae, 350
Cornus
  glabrata, 350
  stolonifera var californica,
    350
  C. x californica, a hybrid
Crepis
  monticola, 362
  occidentalis ssp pumila, 361
Crocidium multicaule, 366
Cruciferae, 340
Cryptantha
  Clevelandii, 354
  corollata, 354
  flaccida, 354
  hispidissima, 354
  nemaclada, 354
  nevadensis var rigida, 354
  sparsiflora, 354
  Torreyana var pumila, 354
Cucurbitaceae, 360
Cupressaceae, 329
Cupressus
  Sargentii, 329
  Sargentii var Duttonii, 329
    C. Sargentii
Cuscuta subinclusa, 351
  C. Ceanothi
Cynoglossum grande, 354
Cyperaceae, 331
Cystopteris fragilis, 328
Datisca glomerata, 348
Datiscaceae, 348
Datura meteloides, 357
  D. inoxia
Daucus pusillus, 349
Delphinium
  californicum, 339
  californicum var interius, 339

hesperium, 339
hesperium var seditiosum, 339
  D. Parryi ssp seditiosum
nudicaule, 338
Parryi, 339
Parryi x variegatum, 339
patens, 338
patens x nudicaule, 338
variegatum, 339
Dentaria integrifolia var
  californica, 341
Deschampsia
  danthonioides, 330
  elongata, 330
Descurainia pinnata ssp
  Menziesii, 341
Deweya
  Hartwegii, 349
    Tauschia Hartwegii
  Kelloggii, 349
    Tauschia Kelloggii
Dicentra chrysantha, 340
Dipsacaceae, 360
Dipsacus fullonum, 360
  D. sativus
Disporum Hookeri, 333
Distichlis stricta, 330
  D. spicata var stricta
Dodecatheon
  Hendersonii, 350
  Hendersonii var bernalinum, 350
    D. Clevelandii ssp patulum
Draba unilateralis, 341
  Heterodraba unilateralis
Dryopteris arguta, 328
Dudleya cymosa ssp Setchellii
  Echeveria laxa var Setchellii, 342
  Echeveria laxa var paniculata, 342
Eastwoodia elegans, 362
Echeveria
  laxa var paniculata, 342
    Dudleya cymosa ssp Setchellii
  laxa var Setchellii, 342
    Dudleya cymosa ssp Setchellii
Echinocystis fabacea, 360
  Marah fabaceus
Elatinaceae, 347

Elatine
 brachysperma, 347
 californica, 347
Eleocharis
 acicularis, 331
 mamillata, 331
  E. macrostachya
Elymus
 condensatus, 330
 glaucus, 330
 glaucus var Jepsoni, 330
 triticoides, 330
Emmenanthe
 penduliflora, 353
 penduliflora var rosea, 353
  E. rosea
Epilobium
 californicum, 367
  E. adenocaulon var Parishii
 minutum, 348
 paniculatum, 348
Epipactis gigantea, 333
Equisetaceae, 328
Equisetum
 arvense, 328
 laevigatum, 328
Eremalche Parryi
 Malvastrum Parryi, 347
Eremocarpus setigerus, 346
Eriastrum
 Abramsii
  Navarretia Abramsii, 351
 filifolium
  Hugelia filifolia, 351
 pluriflorum
  Hugelia pluriflora, 351
Ericaceae, 350
Ericameria arborescens, 362
 Haplopappus arborescens
Erigeron
 miser, 363
 philadelphicus, 363
Eriodictyon californicum, 354
Eriogonum
 angulosum, 335
 Covilleanum, 335
 fasciculatum var foliolosum, 335

 hirtiflorum, 367
 inerme, 335
 nudum, 335
 nudum var auriculatum, 335
 saxatile, 335
 umbellatum var stellatum, 335
 vimineum, 335
 virgatum, 335
  E. roseum
 Wrightii, 335
  E. Wrightii var trachygonum
Eriophyllum
 confertiflorum, 365
 Jepsonii, 365
 lanatum var achillaeoides, 365
 Wallacei, 365
Erodium
 Botrys, 346
 cicutarium, 346
 macrophyllum, 346
 moschatum, 346
Eryngium Vaseyi var castrense, 349
  E. Vaseyi var vallicola
Erysimum capitatum, 341
Eschscholtzia (Eschscholzia)
 caespitosa var hypecoides, 340
  E. caespitosa
 caespitosa var rhombipetala, 340
  E. rhombipetala
 californica, 340
Eucrypta chrysanthemifolia, 352
Euphorbia
 crenulata, 346
 dictyosperma, 346
  E. spathulata
 ocellata, 346
 serpyllifolia, 346
Euphorbiaceae, 346
Evax sparsiflora, 363
Fagaceae, 333
Festuca
 confusa, 330
 dertonensis, 330
 Eastwoodae, 330
 Elmeri, 330
 Grayi, 330
 megalura, 330

myuros, 366
occidentalis, 330
octoflora, 329
pacifica, 330
reflexa, 330
Filago californica, 363
Forestiera neomexicana, 350
Fraxinus dipetala, 350
Fritillaria
agrestis, 333
falcata, 333
lanceolata, 333
Fumariaceae, 340
Galium
Andrewsii, 360
Aparine, 360
Nuttallii, 360
Garrya
Congdoni, 350
Fremontii, 350
Garryaceae, 350
Gastridium ventricosum, 331
Gentianaceae, 350
Geraniaceae, 346
Geranium
carolinianum, 346
dissectum, 346
Gilia
achilleaefolia, 351
gilioides, 352
Allophyllum gilioides
millefoliata, 352
multicaulis, 352
G. achilleaefolia ssp multicaulis
tricolor, 351
sp., 352
Githopsis specularioides, 360
Glinus lotoides, 336
Glycyrrhiza lepidota, 345
Gnaphalium
californicum, 363
chilense, 363
palustre, 363
Godetia
amoena, 348
Clarkia rubicunda
epilobioides, 348

Clarkia modesta
quadrivulnera, 348
Clarkia purpurea ssp
quadrivulnera
quadrivulnera var Elmeri, 348
Clarkia purpurea
Gramineae, 329
Grindelia
camporum var Davyi, 362
G. camporum
camporum var parviflora, 362
rubricaulis, 362
G. hirsutula ssp rubricaulis
Gutierrezia californica, 362
G. bracteata
Habenaria
Michaeli, 333
H. elegans
unalascensis, 333
Haplopappus
arborescens
Ericameria arborescens, 362
linearifolius
Stenotopsis linearifolia, 362
Helenium puberulum, 365
Helianthella californica, 363
Helianthus californicus, 363
Heliotropium curassavicum var
oculatum, 354
Hemizonella minima, 364
Madia minima
Hemizonia
congesta ssp luzulaefolia, 364
H. luzulaefolia
fasciculata, 364
Fitchii, 364
Heermannii, 364
Holocarpha Heermannii
Kelloggii, 364
obconica, 364
Holocarpha obconica
plumosa, 364
Blepharizonia plumosa
pungens ssp interior, 364
H. pungens
virgata, 364
Holocarpha virgata

Herniaria cinerea, 338
Hesperolinon
californicum
Linum californicum, 345
Clevelandii
Linum Clevelandii, 345
micranthum
Linum micranthum, 345
spergulinum
Linum spergulinum, 345
Heterocodon rariflorum, 360
Heterodraba unilateralis
Draba unilateralis, 341
Heteromeles arbutifolia
Photinia arbutifolia, 343
Heuchera micrantha var
pacifica, 342
Hieracium albiflorum, 361
Hippocastanaceae
Sapindaceae, 346
Holocarpha
Heermannii
Hemizonia Heermannii, 364
obconica
Hemizonia obconica, 364
virgata
Hemizonia virgata, 364
Holodiscus discolor, 343
Holozonia filipes, 365
Hordeum
gussoneanum, 330
H. geniculatum
jubatum, 330
murinum, 330
H. leporinum
nodosum, 330
H. brachyantherum
Hugelia
filifolia, 351
Eriastrum filifolium
pluriflora, 351
Eriastrum pluriflorum
Hulsea heterochroma, 365
Hutchinsia procumbens
Capsella procumbens, 341
Hydrophyllaceae, 352
Hydrophyllum occidentale, 352

Hypochoeris glabra, 361
Idahoa scapigera
Platyspermum scapigerum, 341
Iridaceae, 333
Iris
longipetala, 333
macrosiphon, 333
Isoetaceae, 328
Isoetes Howellii, 328
Isopyrum
occidentale, 338
stipitatum, 338
Juglandaceae, 334
Juglans Hindsii, 334
Juncaceae, 331
Juncaginaceae, 329
Lilaeaceae
Juncus
balticus, 331
bufonius, 331
effusus var pacificus, 331
ensifolius, 331
occidentalis, 331
J. tenuis var congestus
oxymeris, 331
patens, 331
sphaerocarpus, 331
xiphioides, 331
Juniperus californica, 329
Koeleria cristata, 330
K. macrantha
Labiatae, 355
Lactuca saligna, 361
Lagophylla
ramosissima, 365
ramosissima ssp congesta, 365
L. congesta
Lamarckia aurea, 330
Lamium amplexicaule, 357
Lasthenia
chrysostoma
Baeria chrysostoma, 365
chrysostoma ssp gracilis
Baeria chrysostama var
gracilis, 365
glaberrima, 365
glabrata, 365

microglossa
  Baeria microglossa, 365
  Pentachaeta laxa, 362
minor
  Baeria uliginosa, 365
Lathyrus Bolanderi ssp
  quercetorum, 345
  L. vestitus
Lauraceae, 339
Layia
  chrysanthemoides, 364
  gaillardioides, 364
  hieracioides, 364
  platyglossa, 364
Leguminosae, 343
Lemmonia californica, 353
Lemna minor, 331
Lemnaceae, 331
Lepidium
  latipes, 341
  nitidum, 341
Lepidospartum squamatum, 366
Leptotaenia californica, 349
  Lomatium californicum
Lessingia
  germanorum var parvula, 362
  hololeuca, 362
  nemaclada, 362
    L. nemaclada var mendocina
Lewisia rediviva, 336
Libocedrus decurrens, 329
  Calocedrus decurrens
Lilaea subulata, 329
  L. scilloides
Lilaeaceae
  Juncaginaceae, 329
Liliaceae, 331
Limnanthaceae, 346
Limnanthes Douglasii, 346
Linaceae, 345
Linanthus
  ambiguus, 352
  androsaceus, 352
  bicolor, 352
  ciliatus, 352
  densiflorus, 352
    L. grandiflorus

dichotomus, 352
parviflorus, 352
  L. androsaceus ssp luteus
pharnaceoides, 352
  L. liniflorus ssp pharnaceoides
pygmaeus, 352
  L. pygmaeus ssp continentalis
Linaria texana, 357
  L. canadensis var texana
Linum
  californicum, 345
    Hesperolinon californicum
  Clevelandii, 345
    Hesperolinon Clevelandii
  micranthum, 345
    Hesperolinon micranthum
  spergulinum, 345
    Hesperolinon spergulinum
Lithophragma
  affinis, 342
    L. affine
  Cymbalaria, 342
  heterophylla var scabrella, 342
    L. Bolanderi
Loasaceae, 347
Lobeliaceae, 361
Loeflingia squarrosa, 338
Lolium multiflorum, 330
Lomatium
  californicum
    Leptotaenia californica, 349
  caruifolium, 349
  dasycarpum, 349
  macrocarpum, 349
  nudicaule, 349
  utriculatum, 349
Lonicera
  hispidula, 360
    L. hispidula var vacillans
  interrupta, 360
  Johnstonii, 360
    L. subspicata var Johnstonii
Loranthaceae, 334
Lotus
  americanus, 345
    L. Purshianus
  crassifolius

humistratus, 345
micranthus, 345
rubriflorus, 345
scoparius, 345
strigosus, 345
subpinnatus, 345
Lupinus
albifrons, 343
bicolor, 343
densiflorus, 344
formosus, 343
microcarpus, 344
    L. subvexus
nanus, 367
rivularis, 343
    L. latifolius
succulentus, 343
Luzula multiflora, 331
    L. comosa
Lythraceae, 348
Lythrum adsurgens, 348
Madia
elegans, 364
exigua, 364
madioides, 364
    minima
        Hemizonella minima, 364
radiata, 364
sativa ssp dissitiflora, 364
    M. gracilis
Malacothamnus Fremontii ssp
    cercophorus
    Malvastrum Fremontii var
        cercophorum, 347
Malacothrix
Clevelandii, 361
Coulteri, 361
floccifera, 361
Malus sylvestris
    Pyrus Malus, 343
Malva parviflora, 367
Malvaceae, 347
Malvastrum
Fremontii var cercophorum, 347
    Malacothamnus Fremontii ssp
        cercophorus
Parryi, 347

Eremalche Parryi
Marah fabaceus
    Echinocystis fabacea, 360
Marrubium vulgare, 355
Marsilea vestita, 328
    M. mucronata
Marsileaceae, 328
Matricaria matricarioides, 366
Medicago
apiculata, 344
    M. polymorpha var brevispina
hispida, 344
    M. polymorpha
lupulina, 344
Melica
californica, 330
imperfecta var flexuosa, 330
Torreyana, 330
Melilotus alba, 344
    M. albus
Mentzelia
dispersa, 348
gracilenta, 348
laevicaulis, 347
Lindleyi, 347
micrantha, 348
Micropus californicus, 363
Microseris
acuminata, 361
Douglasii, 361
Lindleyi, 361
linearifolia, 361
    M. linearifolia
sylvatica, 361
tenella var aphantocarpha, 361
    M. Douglasii ssp tenella
Microsteris gracilis
    Phlox gracilis, 351
Mimetanthe pilosa, 358
    Mimulus pilosus
Mimulus
androsaceus, 358
aurantiacus, 358
Bolanderi var brachydontus, 358
    M. Bolanderi
cardinalis, 358
floribundus, 358

nasutus, 358
pilosus
  Mimetanthe pilosa, 358
Monardella
  Breweri, 357
  Douglasii, 357
  villosa ssp subserrata, 357
Monolopia
  lanceolata, 365
  major, 365
Montia
  fontana, 336
    M. Hallii
  gypsophiloides
    Claytonia gypsophiloides, 336
  perfoliata
    Claytonia perfoliata, 336
Muilla serotina, 332
  M. maritima
Myosurus
  lepturus, 339
    M. minimus
  minimus, 339
  minimus var filiformis, 339
Najadaceae, 329
  Zannichelliaceae
Navarretia
  Abramsii, 351
    Eriastrum Abramsii
  intertexta, 351
  mellita, 351
  pubescens, 351
Nemacladus
  montanus, 361
  ramosissimus, 361
Nemophila
  heterophylla var nemorensis, 352
    N. heterophylla
  Menziesii, 352
  pedunculata, 352
  sepulta, 352
    N. pedunculata
Nicotiana
  attenuata, 357
  glauca, 357
Oenothera
  contorta var strigulosa, 348

  O. dentata
decorticans, 348
  O. Boothii ssp decorticans
deltoides var cognata, 348
micrantha var Jonesii, 348
  O. hirtella
Oleaceae, 350
Onagraceae, 348
Orchidaceae, 333
Orobanchaceae, 359
Orobanche
  fasciculata, 359
  tuberosa, 367
    O. bulbosa
  uniflora var Sedi, 359
Orthocarpus
  attenuatus, 359
  densiflorus, 359
  erianthus, 359
  erianthus var micranthus, 359
  purpurascens, 359
  pusillus, 359
Osmaronia cerasiformis, 343
Osmorhiza
  brachypoda, 349
  nuda, 349
    O. chilensis
  occidentalis, 367
Paeonia Brownii, 338
Panicum capillare var
  occidentale, 331
Papaver heterophyllum, 340
  Stylomecon heterophylla
Papaveraceae, 340
Parvisedum pentandrum, 341
  Sedella pentandra, 341
Pectocarya
  linearis var ferocula, 354
  penicillata, 354
  pusilla, 354
  setosa, 354
Pedicularis densiflora, 358
Pellaea
  andromedaefolia, 328
  mucronata, 328
Penstemon
  breviflorus, 357

corymbosus, 357
heterophyllus, 357
heterophyllus ssp Purdyi, 357
Pentachaeta
exilis, 362
Chaetopappa exilis
laxa, 362
Lasthenia microglossa
Perideridia
californica, 349
Gairdneri, 349
Petunia parviflora, 357
Phacelia
Breweri, 353
californica var imbricata, 352
P. imbricata
ciliata, 353
distans, 353
divaricata, 353
Douglasii var petrophila, 353
Fremontii, 353
nemoralis, 367
phacelioides, 353
ramosissima var suffrutescens, 353
Rattanii, 353
stimulans, 353
P. imbricata
tanacetifolia, 353
Phlox gracilis, 351
Microsteris gracilis
Pholistoma
aurita, 352
P. auritum
membranacea, 352
P. membranaceum
Phoradendron
flavescens var macrophyllum, 334
P. tomentosum ssp macrophyllum
villosum, 334
Photinia arbutifolia, 343
Heteromeles arbutifolia
Phragmites communis, 330
P. communis var Berlandieri
Pickeringia montana, 343
Pinaceae, 328
Pinus

Coulteri, 329
ponderosa, 328
Sabiniana, 329
Pityrogramma triangularis, 328
Plagiobothrys
acanthocarpus, 355
arizonicus, 355
bracteatus, 354
canescens, 355
infectivus, 355
myosotoides, 355
nothofulvus, 355
tenellus, 355
Plantaginaceae, 360
Plantago
Hookeriana var californica, 360
major, 360
Platanaceae, 343
Platanus racemosa, 343
Platyspermum scapigerum, 341
Idahoa scapigera
Platystemon californicus, 340
Plectritis
ciliosa, 360
ciliosa ssp insignis
Aligera rubens, 360
macrocera, 360
Aligera collina, 360
magna, 360
P. congesta ssp brachystemon
samolifolia, 360
P. congesta ssp brachystemon
Poa
annua, 330
Howellii, 330
P. Bolanderi ssp Howellii
pratensis, 330
scabrella, 330
secunda, 330
scabrella
Pogogyne serpylloides, 356
Polemoniaceae, 351
Polygonaceae, 334
Polygonum
aviculare, 367
Parryi, 334
Polypodiaceae, 328

Polypodium vulgare var
  Kaulfussii, 328
  P. californicum
Polypogon
  lutosus, 331
    P. interruptus
  monspeliensis, 331
Polystichum munitum, 328
Populus
  Fremontii, 333
  trichocarpa, 333
Portulacaceae, 336
Potentilla glandulosa, 343
Primulaceae, 350
Prosopis chilensis, 343
    P. glandulosa var Torreyana
Prunus
  emarginata, 343
  ilicifolia, 343
  subcordata, 343
  virginiana var demissa, 343
Psilocarphus tenellus, 363
Psoralea
  californica, 345
  macrostachya, 345
  physodes, 345
Pterostegia drymarioides, 334
Puccinellia simplex, 329
Pyrus Malus, 343
    Malus sylvestris
Quercus
  agrifolia, 334
  chrysolepis, 334
  Douglasii, 334
  dumosa, 334
  durata, 334
  Garryana, 333
  Kelloggii, 334
  lobata, 333
  morehus, 334
  Wislizenii, 334
Radicula
  curvisiliqua, 341
    Rorippa curvisiliqua
  Nasturtium-aquaticum, 341
    Rorippa Nasturtium-aquaticum
Rafinesquia californica, 361

Ranunculaceae, 338
Ranunculus
  californicus, 339
  hebecarpus, 339
  trichophyllus, 339
    R. aquatilis var capillaceus
  trichophyllus var hispidulus, 339
    R. aquatilis var hispidulus
Rhamnaceae, 346
Rhamnus
  californica ssp tomentella, 346
  crocea ssp ilicifolia, 346
Rhus diversiloba
    Toxicodendron diversilobum, 346
Ribes
  amarum, 342
  aureum var gracillimum, 342
  californicum, 342
  malvaceum, 342
  quercetorum, 342
  sanquineum var glutinosum, 342
  speciosum, 342
Rigiopappus leptocladus, 365
Rorippa
  curvisiliqua
    Radicula curvisiliqua, 341
  Nasturtium-aquaticum
    Radicula Nasturtium-
      aquaticum, 341
Rosa
  californica, 343
  gymnocarpa, 343
Rosaceae, 343
Rubiaceae, 360
Rubus vitifolius, 343
Rumex
  Acetosella, 334
  conglomeratus, 334
  crispus, 334
  salicifolius, 334
Sagina
  apetala var barbata, 336
  occidentalis, 336
Salicaceae, 333
Salix
  Breweri, 333
  Hindsiana, 333

laevigata, 333
laevigata var araquipa, 333
lasiolepis, 333
melanopsis, 333
Salsola Kali var tenuifolia, 336
    S. pestifera
Salvia
    carduacea, 355
    Columbariae, 355
    mellifera, 355
Sambucus caerulea, 360
Sanicula
    bipinnata, 349
    bipinnatifida, 349
    crassicaulis, 349
    saxatilis, 349
    tuberosa, 349
Sapindaceae, 346
    Hippocastanaceae
Satureja Douglasii, 356
Saxifraga californica, 342
Saxifragaceae, 342
Scribneria Bolanderi, 330
Scrophularia californica, 357
Scrophulariaceae, 357
Scutellaria
    siphocampyloides, 355
    tuberosa, 355
Sedella pentandra, 341
    Parvisedum pentandrum
Sedum
    radiatum, 342
        S. stenopetalum ssp radiatum
    spathulifolium, 342
Selaginella Bigelovii, 328
Selaginellaceae, 328
Senecio
    aronicoides, 366
    Breweri, 366
    Douglasii, 366
    vulgaris, 366
Sidalcea
    diploscypha, 347
    malvaeflora, 347
Silene gallica, 338
Sisyrinchium bellum, 333
Sitanion

Hanseni, 330
hystrix, 330
jubatum, 330
Smilacina
    amplexicaulis, 333
        S. racemosa var amplexicaulis
    sessilifolia, 333
        S. stellata var sessilifolia
Solanaceae, 357
Solanum umbelliferum, 357
Solidago californica, 362
Sonchus
    asper, 361
    oleraceus, 367
Spergularia
    atrosperma, 338
    salina, 338
        S. marina
Stachys
    ajugoides, 357
    pycnantha, 357
    rigida ssp quercetorum, 357
Stellaria
    media, 336
    nitens, 336
Stenotopsis linearifolia, 362
    Haplopappus linearifolius
Stephanomeria
    exiqua var coronaria, 361
    virgata, 361
Stipa
    lepida, 331
    pulchra, 331
Streptanthus
    albidus, 340
    Breweri, 340
    callistus, 340
    Coulteri var Lemmonii, 340
    glandulosus, 340
    lilacinus, 340
        Thelypodium flavescens
Stylocline
    filaginea, 363
    gnaphalioides, 363
Stylomecon heterophylla
    Papaver heterophyllum, 340
Symphoricarpos

albus, 360
  S. rivularis
mollis, 360
Tauschia
  Hartwegii
    Deweya Hartwegii, 349
  Kelloggii
    Deweya Kelloggii, 349
Taxaceae, 329
Thalictrum polycarpum, 339
Thelypodium
  lasiophyllum, 340
  flavescens, 340
Thermopsis macrophylla, 343
Thysanocarpus
  curvipes, 341
  laciniatus var crenatus, 341
  radians, 341
Tillaea
  aquatica var Drummondii, 341
  erecta, 341
Tonella tenella, 357
  Torreya californica, 329
Toxicodendron diversilobum, 346
  Rhus diversiloba
Trichostema lanceolatum, 355
Trifolium
  albopurpureum, 344
  amplectens, 344
  barbigerum, 344
  bifidum, 344
  ciliolatum, 344
  cyathiferum, 344
  depauperatum, 344
  dichotomum, 344
  dichotomum var turbinatum, 344
  fucatum, 344
  gracilentum, 344
  Macraei, 344
  microcephalum, 344
  microdon, 344
  obtusiflorum, 344
  oliganthum, 344
  olivaceum var griseum, 344
  tridentatum, 344
  variegatum, 344
  Wormskjoldii, 344

Trillium sessile var giganteum, 333
  T. chloropetalum var giganteum
Tropidocarpum
  capparideum, 341
  gracile, 341
Umbelliferae, 349
Umbellularia californica, 339
Urtica
  gracilis var holosericea, 334
    U. dioica ssp gracilis var
      holosericea
  urens, 334
Urticaceae, 334
Valerianaceae, 360
Verbascum thapsus, 357
Verbena prostata, 355
  V. lasiostachys
Verbenaceae, 355
Veronica
  americana, 358
  peregrina ssp xalapensis, 358
Vicia
  americana var truncata, 345
  californica, 345
  sativa, 345
Viola
  Douglasii, 347
  pedunculata, 347
  purpurea, 347
    V. quercetorum
  Sheltonii, 347
Violaceae, 347
Vitaceae, 347
Vitis californica, 347
Woodwardia fimbriata, 328
Wyethia
  angustifolia, 363
  helenioides, 363
Xanthium spinosum, 365
Zannichelliaceae
  Najadaceae, 329
Zannichellia palustris, 329
Zauschneria californica, 348
  Z. californica ssp mexicana
Zygadenus
  Fremontii, 331
  venenosus, 331